PLANT, WILDLIFE AND MAN

PLANT, WILDLIFE AND MAN

–A Conservation Scenario of Bay Islands and the Himalayas–

by

A.B. Chaudhuri, *IFS (Retd.)*

Ex-Director, Forest Survey of India

and

Consultant, Environment and Forests

ASHISH PUBLISHING HOUSE

5 Ansari Road, Darya Ganj

New Delhi- 110 002

Published by :
S.B. Nangia
for **Ashish Publishing House**
8/81, Punjabi Bagh
New Delhi-110 026

Tel : 50 05 81
541 09 24

Show Room : 5 Ansari Road, Darya Ganj,
New Delhi-110 002

Tel : 3274050

ISBN 81-7024-464-1

1992

Printed at :
EFFICIENT OFFSET PRINTERS
215, Shahzada Bagh Indl. Complex,
Phase-II, Phone : 533736, 533762
Delhi - 110035

DEDICATED TO THE MEMORY OF

LATE V.S. RAO OF INDIAN FOREST SERVICE,

A DEDICATED FORESTER AND A NATURE LOVER,

WHO INITIATED ME TO WRITING ON FORESTS

AND NATURE

FOREWORD

In the recent past serious concern has been expressed over the depletion of forests and wildlife in the country. The need for improving the tree-cover is being increasingly highlighted by the erratic behaviour of the monsoon, severe soil erosion leading to decline of productivity, siltation in rivers etc. It is being realised that the destiny of our country is in inextricably linked with conservation of floral and faunal resources. The on-going social forestry programmes all over the country and wildlife conservation schemes in many areas are some of the manifestations of this understanding.

In this context Shri A.B. Chaudhuri's work on flora and fauna and their actions and interactions in the life-support system has been timely and appropriate. Shri Chaudhuri has presented in a lucid style a wealth of complex scientific observations. He has succeeded in bringing out his long experience in the Indian Forest Service in the present volume.

I hope the book would stimulate its readers to further studies on flora and fauna in the country and be useful to school and college students, educationists, conservationists and administrators.

R.N. De

Secretary,
Forest Department
Government of West Bengal
Writer's Biulding
Calcutta

PREFACE

The author's intention in presenting diverse facts of Indian environment in this book is partly contained in a few lines mentioned in 'The World Wildlife Charter, 1962' which reads -

"Love and respect for living creatures must always be seen to flow from love and respect for the highest qualities and aspirations of mankind. All who see the need for saving the world's wildlife whether they are animal lovers, naturalists, scientists, sportsman or simply people of good will and public spirit, must endeavour to work together amicably and to good purpose. United we can and will save the world's wildlife, divided we could lose it for ever".

But the author extends his arena of studies to cover trees, shrubs, herbs, their endemism, status, conservation effort and very many other environmental aspects including human interactions. Man's life on earth is intricately interwoven with animals, birds and fish, insects and beasts, worms and plants and all these have been discussed in brief in Indian context.

India has inherited an incredibly beautiful and complex set of plants and animals, but for many years sincere efforts had not been made and the simplest principals of conservation had been neglected.

Architectural monuments are great work of mankind's wealth. And so are the plants and animals. The works of arts adorn and enrich life. And man's world would also be a wasteland without flowers and trees, birds and beasts. "Nature excites and attracts inquisitive minds with the beauty and variety of forms, with their grandeur, with extreme power, mysteriousness and strict consistency of its phenomena. ...." (V.O. Kovalevsky). The book takes care of all these qualities of nature.

Of the four sections, section'A' deals with various environmental values of trees, shrubs, herbs, flowers, grasses and mangroves, rhythm of tropical forests and life systems of aquatic habitat; besides, it describes the behaviour patterns of man-eaters and rogues in killing human beings and the environmental hazards faced by tribals. An elaborate picture of wildlife potentialities of India, existing national parks, wildlife sancturies, has been presented in section B, to highlight the fact that India has a large extent of conservation areas where thousands of tired nerve-shaken, civilized people are beginning to realise that going to such areas in going home; man needs wilderness areas and mountain parks to meet his various needs. Section 'C' deals with faunal and floral wealth of Andaman and Nicobar Islands and various problems facing the tropical forests of the islands. These tropical forests are the repository of unbelievable biological wealth and diversity. Various evolutionary and biogeographic factors have created a unique and highly vulnerable fauna in these oceanic islands with a very high degree of endemism. Through long, years of isolation the original faunal population of the island have evolved and specialised to such an extent that most of them are vulnerable to any special modifications. The great Nicobar island is of unique importance from environmental point of view. It contains a sizeable untouched lowland evergreen forests which is an excellent example of an ecosystem in which climate, soil, vegetation, and fauna are components in an extremely complex equilibrium. An up-to-date information has been presented on wildlife conservation areas. Section 'D' deals with unique biogeographical position of Sikkim and adjoining areas. A threadbare discussion has been made on various floral and faunal resources of the area in different altitudinal zones.

It has been desired to have a meaningful book full of profound meaning which the people of various walks of life will usefully utilise. It is hoped the book will also serve as a reference book besides meeting demands of a text book in the schools and colleges and to the environmentalists. Various information compiled in the book form a dark mine which conceals immense treasures of information in its profound depth.

131 N.S.C. Bose Road, A.B. Chaudhuri
Block - 10 Flat-4
Calcutta

India, the land with vast natural resources, diverse flora and fauna and a heavy population

CONTENTS

SECTION A

Plants, Animals and Habitat Conservation for People

SECTION B

India Saves Its Varied Wildlife Ecosystem

SECTION C

A Wisdom to Conserve Natural Resources of Andaman and Nicobar Islands

SECTION D

Sikkim Himalayas The Nucleus of Natural Resources of Great Himalayan Range

SECTION A

PLANTS, ANIMALS AND HABITAT CONSERVATION FOR PEOPLE

1

THE TRINITY OF ENVIRONMENT
(TREES, SHRUBS AND HERBS)

The principal lifeforms of vegetation, the trees, shrubs and herbs are alpha and omega of our environment in India. Broadly, vegetation, soil, air, micro and macro fauna form our environment; but of all these components, the vegetation play a major role in stabilizing the structural configuration of nature. Rightly, the trees, shrubs and the herbs may be called the Trinity of our environment as they collectively play a major role in modulating the structure.

The nature's vast inexhaustible renewable resources should be managed and conserved; man has to derive from them the recreational, aesthetic, ethical, ecological and commercial and scientific values. Trees, shrubs and herbs form a part of the forest community or forest biocoenosis. Each of these lifeforms individually play a specific role but collectively as forests they play unique conservation roles to fight various evil forces of nature.

Trees, according to a pragmatic definition of a forester, are perennial woody plants that reach a minimum height of 3 to 4 metres, have a well formed crown of foliage and attain a diameter of 7.5 at 1.35 metres from ground level. Trees are natural and perennial lend marks and sacred memorials; some are famous and have a special religious, aesthetic and sentimental associations. They keep our hearths aglow on the cold winter nights and screen us from the scorching sun; they form the beams of our house, boards of our table, bed on which we lie, timber of our boats, doors of our homesteds, the wood of our cradles and the shell of our coffins.

Men plant trees for posterity. In planting a tree men lays the foundation for a ship that crosses the sea, must that carries the sail, a planks that fights the gale and a rafter that holds the house. A lone tree cannot brave the blast of a storm but a compact mass of trees can defy a storm.

Trees that constitutes forests are part of our amazing resources that have contributed to the prosperity of our nation. They are also sources of history and beauty, gifts of God and friends of man. John Elvin, the English patriot and philosopher, as early as 1664 warned and advised, "Men seldom plant trees till they begin to be wise, that is, till they grow old and find by experience, the prudence and necessity it". When trees grow collectively–a forest is born; an environment is created and plant and animal succession follows.

Trees interrupt reflected light, screen out undesirable sights, border the cities, lakes and streams. Trees with the rough, thick and sticky leaf surface collect soot and oil and reduce environment pollution. Trees adorn our homesteds and our cities, shelter our wildlife; they act as soothing balm on the wearied and harried city dwellers. Trees have played an important role in the evolution of huminity; they are greatest benefactors, friends, teachers and preceptors. They have the sense of beauty and instincts of love, service, sacrifice and harmony.

Forests are for human survival. "An age builds up cities, an hour destroys them; in a moment the ashes are made but forest is a long time growing". Forests are protective, productive and aesthetic. The vital impacts of forests on climate, soil erosion and floods are acceptable to all. They protect us against winds, storms and rains. Where forests abound, the rains are plentiful. Forest soil with its microbes and vegetative cover acts as a natural filter.

Forests are perennial source of raw materials necessary for a variety of industries–pulp and paper, panel products, match wood, sports goods, plywood, fibre board, saw mills, constructional wood, etc.

Tropical forests are dark and impenetrable and abound with insects, reptiles, amphibians and mammals. These forests throw a baffling and intriguing challenge in their multistoreyed structure which is diverse and complex.

India has more than 1500 tree species of which only few are regenerating. Overexploitation, selective exploitation, grazing and fire are threatening their survival. Trees which occupy the topmost layer of a forest may not serve a foolproof function against various atmospheric hazards; it also needs a well knit shrub and herb layer. Of the three vertical layers, top middle and bottom, the middle layer has the maximum number of species; next is the bottom layer and it is minimum in the top layer. It is needless to mention that there is hardly much forest left in the country where the bottom and middle storey trees still stand. Even the top layers have been stripped off of many tree species. All these give a dismal picture of the state of vegetation in this country.

Extensive areas supporting natural tree growth have come under reckless exploitation and eventual intense environmental degradation. Deforestation of 5 million hectares of land added by ravine affected 7 million and another 48 million hectares lost in river valley projects offer a gloomy picture of tree cover of the country. The much bloated claim that India has 22.8 percent of it's geographical area under forests is far from true. India has hardly 6.8 percent of it's geographical area under somewhat good vegetation cover; more than 13.4 percent of the geographical area is under poor cover. Even than the situation has turned worse owing to uneven distribution of the forests. Vast erodible areas in the hilly tracts of western ghat, Himalayas and central India have been denuded beyond immediate recuperation. All these leave one to think that the tree vegetation in India are now in a bad shape to offer any viable safeguards against the environmental hazards.

Shrubs

Shrubs are low woody plants or a bush with no trunks, no crown of foliage, often multistemmed and usually of a height less than 3 to 4 metres. They are mostly shadow bearer and form an essential component of lifeform of vegetation community.

Various species of shrub with their intimate assemblages and forms are biologically more important. They provide an ideal cover to the terrestial animals. The herbiviroes can graze comfortably on the fodder shrub because of their identical height. In a dense tree forest the shrubby layer contains about 50 percent of regeneration of tree species.

The degree and duration of the shade have a positive bearing on the composition and density of shrub crop. Heavy shade results in a very limited number of shrub flora. The intensity of light is concerned with photosynthesis and food production; it also affects flower and fruit production. Specific insect and avifaunal species are concerned with shrubs for food and pollination. At least eight shrub genera form an imperative source of larvae food of the moths and butterflies.

Over-exploitation of tree crops in all the three storeys cause considerable physical damage to shrubs; these shrubs are essential for soil binding and create a micro climate for the survival of young plant regeneration; besides, they help litter formation and conservation of water.

Shrubs and young regeneration of other plant species suffer irreparable losses due to repeated grazing and fire. These latter two factors are directly responsible for eradication of plant species where only those with underground stem survive. The author's research reveals that only about 40 to 50 percent of the species survive such depredation; also the density of occurrence of species is affected. Vast tracts of forests where grazing and fire have a free play have now very coarse grass and a few fire hardy shrub species.

About ninety species of shrubs are conspicuous in the temperate hills of India which have colourful flowers; they also grow in profusion. Special mention may be made of rhododendron and roses. In the plains, however, more than one hundred species occur but they do not have conspicuous flowers. In eastern India there are at least a hundred species of fodder shrub; it is about two hundred in other parts of the country. There are at least forty genera of shrubs yielding edible fruits which attract avifauna, animals and men alike.

A survey made by the author all over the country has revealed that only twenty five species of shrubs of the genera, *Lantana, Eupatorium, Clerodendrum, Calotropis, Cassia, Carissa, Capparis,* etc. are found conspicuous and are occurring in profusion all over the country. It should be considered an ominous indication of systematic disappearance of species due to primarily anthropogenic factors. The number of shrub species in the country is about 4500 (tree species about 1500) and wide occurrence of only 25 species only is hardly to be believed, but it is fact believe it or not.

Herbs

Of all lifeforms of plants, the number of herb species outnumber others. Roughly, the number of shrub species is about double the trees, while the number of herbs is about four times.

Herb has been defined as "Plants with no woody stems above the ground." Others define it as "A plant the stem of which dies every year" or as, "A plant with no persistent parts above the ground, as distinct from shrubs and trees". The concise Oxford dictionary defines it as, "A plant of which the stem is not woody or persistent and which dies down after flowering; plants of which leaves etc. are used for food, medicine and flavour, etc." It has been defined in many other books ranking it with weeds which are not valued for beauty, that grow wild and hinder and growth of superior species. It has found to grow where it is not desired and where it grows luxuriantly and plentifully.

But no plants are weeds. Each species, has a part to play, either singly or as a member of the community. Each maintains a mutual harmony in it's niche and each niche plays a positive role in moulding the environment. Some herbs may be troublesome, most aggressive, harmful or annoying to man or to his agriculture and are terms as weeds. The herbaceous weeds grow luxuriantly; they are resistant, have a high reproductive capacity, gregarious, have a high level of seed dormancy, are quick growing, often poisonous, deep rooted, annual or perennial and abundantly seed bearing. Several herbs are parasitic, some are hosts of fungus and bacterial diseases, some are hosts of nematodes and some are associated with insects. Many are delicious food for animals and man alike; many are useful as medicinal plants. Some weeds are guardians of soil. They provide protective cover against erosive action of rain drops and run-off. They are also used as mulch in cultivated land.

The herbs come in profusion in swamps, marshes and on other sites which accumulate water in rains. In permanent ponds, weeds also predominate. Aquatic life burst into activities and complete their life cycle within a short span. Succession of plants and animals goes on simultaneously in vertical and horizontal strata. Herbs--submerged, floating, emergent--grow in profusion and have diversified animal periphyton of insects, amphipods, mites, snails, etc. The floral and faunal complex attract aquatic birds. The ponds, marshes and swamps become roosting sites and feeding ground for many species of birds. Heron feed on shallow water, grebe, cormorants and terns feed on fish in open water, the egrets bitterns eat fish in shallow water, cranes and coots are omnivorous, rails, ibises, stilts and snipes probe around in mud, but various ducks eat seeds, roots and soft parts of weeds. In India the aquatic flora consist of 171 species of plants belonging to 90 genera and 39 families.

Besides the aquatic sites, herbs also grow on road side, rice field, dry cultivated land and in various types of forests. In spite of their wide range of occurrences relentless biotic pressure, flood and erosion have reduced the number of species and many have been threatened to the verge of extinction. Over-exploitation of medicinal, decorative and other commercial species has been responsible for shrinkage of many species. Shifting cultivation has also led to the complete eradication of many species. Many herbs have also been over-exploited for fodder purposes.

In fact the environmental role of trees, shrubs and herbs should be studied in conjunction with grasses, sedges, palms, epiphytes, parasites, etc. as all these lifeforms constitute inseparable components of the plant communities to mould the environment.

Herbs form eighty percent of the medicinal plant species. They need meticulous protection. There are many diffused and inconspicuous herbs which have showy flowers. Many herbs grow in grasslands, many in tall reeds and many in alpine grasslands. Innumerable herbs growing in temperate hills have also showy flowers. Many species of canes, yams, lily, asparagus, etc. are being over-exploited. There are many species of aroids besides wild banana, ginger, turmeric, cardamom which grow over extensive areas and form conspicuous physiognomy of the forest floor.

But the enumeration of the essential qualities of herb layer remains incomplete unless some mention are made on climbers, ferns and epiphytes.

Large woody climbers, another lifeform and a biotic association, are remarkable because of their fantastic forms. They add complexities to the vertical structures of forests. There are, however, delicate climbers, shrubby climbers, woody climbers and stranglers apart from root climbers like some climbing palms. There are about six hundred climbers the half of which is shrubby climbers.

Ferns are another type of lifeform which influences the environment. Some are fodder, some provide material for interior decoration, some protect soil and some are essential ingredients for Landscape planning. They grow in tropical and temperate forests both in horizontal and in vertical strata.

Epiphytes, however, are another lifeform that derive support from host plant. Some have remarkable adaptation to hold water. They consume water in a restricted way and many conserve water in their tubers. Base of overlapping leaves of some ferns form a special niche where soil, water and humus are held and nitrogen accumulates. Arboreal ants colonise, lay their eggs and drag seeds that grow in the niche. Insectivorous birds gather to eat ants and eggs; birds also gather and gradually a faunal association grows. Epiphytes are chief components of the vertical structure of our forests.

Thus the song of the trees, shrubs and herbs, the Trinity of environment has been sung in this paper. Much have been written to convince the readers that no amount of administrative and legislative and protective measures to save the country from environmental disaster would be effective and lasting without creation of a dense cover of vegetation all over the country.

2

THE WONDERFUL ENVIRONMENT OF FLOWERS

In the Bhagavat Gita Shri Bhagawan said, "Whosoever offers me with devotion a leaf, a flower, a fruit or water, that offering of love, the pure of heart I accept".

Countless appellations have been used to express the beauty and quality of flowers. Flowers is divine, pure and sacred; it is an emblem of joy, luck and eternal pleasure and is linked with innocence, strength, softness, renunciation and satisfaction. Flower is enjoyment, pleasure, a symbol of universal beauty and sweetness. It is offered in the birth of a child as an emblem of joy; in death it is offered as an expression of sympathy and oneness in adversary. It is also inseparable from a reception ceremony or a farewell congregation. Without flower the life of a man is dull, incomplete and a drud grey. Floral decorations in a woman's coiffure and person are signs of purity and beauty. The fragrance of flower brings a lasting sense of peace and tranquillity in a man and creates a serene atmosphere. In a wedding night, it's colour and fragrance pervades an air of completeness to bring two souls together.

In ancient days flowers and poems were the only gifts exchanged by lovers. Since antiquity, mythology and religion, folklore and legend, magic and superstition were intermingled with flowers. Every human sentiment, joy and grief, triumph and woe, admiration and gratitude, love and desire has it's floral emblem.

Folklore and Symbolism

Countless pages have been written about folklore and symbolism of flowers. Lily is considered as immortal and sacred flower of motherhood. The Sumerians and Babilonians considered it a symbol of fruitfulness. In ancient Rome it was the emblem of Juno, the Goddess of light, marriage, sky and motherhood. In Christian symbolism the Lily represented purity, chastity and innocence and as the symbol of resurrection and Easter; the white Madonna lily was considered as special flower of the Holy virgin. According to ancient semetic legend, the lily sprang from the tears of Eve when expelled from the Garden of Eden. Lotus is the most revered with a deep rooted religious significances. It has an uninterrupted symbolic history of over five thousand years. The Egyptian lotus is dedicated to Horus, the God of the Sun. It was also the symbol of resurrection. In Tibet, Nepal and India it was the emblem of mystery. The Hindus believe that Brahma was born in the sacred bosom of the flower. In India it was considered essential ingredient of the puja of Goddess Durga. It is also interwoven with the life style of the Indians and, therefore, it has been selected as India's national flower. In Persia, the lotus was venerated as the symbol of the sun and of light. Rose is mentioned in every poetic work from the biblical rose of Sharon in the song of Solomon. In mythology the rose is believed to be the symbol of eternal beauty, youth and love. Rose oil, rose water and rose honey were widely used in the past. In India the flower of Datura (thorn apple) is associated with the worship of Lord Shiva, the creator and destroyer of the universe. The worship of Kali is associated with the flower Jaba. Passion flower (Jhumko lata) is the symbol and instruments of the passion; it's ten petals represents ten faithful apostles; the corona symbolises the crown of thorn and five stamens as five wounds. Sunflower is the representative of Sun God. The Iris was considered by the ancient Egyptians as a symbol of power and placed on the brow of the Sphinx. The petals of the iris symbolised faith wisdom and valour. Heliotrope flower is considered a symbol of eternal love and admiration in Greek mythology. Many such examples can be cited.

Flower Colour Scent and Insects

Colouration, shape of flower, scent and the sizes of flowers are related to the biological cycle of various insects and avifaunal species. Avifaunal and insect species are attracted for food and in the process pollinate flowers. Bees are sensitive primarily to yellow colours, but also to red and blue; hive-bee prefers blue and yellow flowers. Bats prefer dreb green colours; beetles and moth prefer red, purple, white and pink flowers of some specialised plant families. Butterflies prefer scarlet, red or yellow flowers. All these principal colours are provided by Himalayan plants. An analytical work on colouration was done by the author on Himalayan flowers where he found that about 53 per cent of plants bear white to yellow coloured flowers; red form about 12 per cent, pink 10 per cent, and other assorted colour about 25 per cent. Taking trees separately white colour covers about 33 per cent (in shrub it is 33 per cent and in herb 24), yellow 23 per cent (in shrub it is 23 percent and in herb 22), purple 4 per cent (in shrub it is 10 per cent and in herb 8), pink 5 per cent (in shrub it is 9 per cent and herb 13), and blue 8 per cent (in shrub it is 6 per cent and herb 15). The balance provide other shades.

The phenomena of colouration of flowers are natures own masters mind of creation. It is in perfect consonance with the nature and pattern of behaviour of various groups of insects and birds. Even the time of bloom of flowers is correlated with the life cycle of insects and mammals. In short, it may be innumerated that night blooming flowers are white and strongly scented. The moths avoid light coloured flowers. Diptera prefers flowers having the smell of fruits; hymenoptera likes the smell of overripe fruits.

Insects and flowers are mutually interrelated in evolution as reciprocal selective factors adopted for each other. Attractive floral pattern and specialised pollinated mechanism evolved in correlation with sensory development of their pollination. Beetles and flies visit elementary floral types unable to operate complicated pollination mechanism. Bees and butterflies are able to pollinate complex flowers. Bumble bees can collect nectar from spurred petals of *Linaria, Delphinium, Acquilegia, Aconitum, Viola* etc. Insect behaviour in relation to flowers is immensely interesting. The bee dances various types of dance such as round dance and waggle–tail dance and similar others. Colour visions are well developed in bees and butterflies. Butterflies prefer to visit flowers with long narrow tubes and flat topped petals; their tongue is dry, smooth and very long and slender. They are more of a nectar collector than pollen. Butterfly flowers are generally scented. Most of the Moth flowers are nocturnal and fragrant. Some diurnal flowers are also associated with moth. Some flowers are partly adapted for day pollination by bees and partly for night pollination by moths. Some *Delphinium, Petunia, Rhododendron* flowers etc. are usually considered to be "bumble bee" or "butterfly" flowers, which do not close after opening and are available for exploitation by crepuscular moths. The stripe hawk moth prefers blue-violet-purple colour range. Positive correlation has been established corolla tube and tongue length of avifaunal species.

Regarding scent, indoloid chemicals imparts smell like decomposing animal matter as in Araceae, Anonaceae, Aristolochiaceae and Asclepiadaceae. Aminoids are sweet smelling, Denzoids smell that of jasmine, primorse and honesuckle. Terpinoidas in *Thymus* and paraffinoid as in *Veleriana.*

Fly flowers smell unpleasantly of sweet, faeces, urine or bad meat (aminoid smell). Next to the bees, flies visit more species of flowers, but they are nectar visitors primarily. Flower scents are volatile essential oils and complex mixtures of compound. The scent of rose is due, to the terpene alcohols alephatic, geraniol, nerol and citronellol and also due to phenols eugenol, methyleuginol and isoengenol. The scent of day flowers differ in kind from those of noctural flowers being less sweet and spicy (bat flowers smell mousy and sour). Some orchids have different types of smell during day and night. Flowers of Labiatae (family of sacred basil) are scentless, while flowers have a higher percentage of scented blooms than any other colour.

India's Floral Resources

In their wide variety of colours, numbers and arrangement of constituent parts and relative positions of plants that produce them, flowers present a seemingly endless variety of combinations. The range from minute blossoms of duckweed to the grotesque giant fleshy bloom of the monster flower of *Reflessia* (exotic species).

India is fortunate in having myriads of flowering plants, many with varieties of colours, both pure and variegated. Vermillion red Palas (the flame of the forest), yellow amaltas, red krishnachura, scarlet semul, mauve and white kanchan are common. Besides, there are innumberable species of showy flowers in the climbers, herbs and shrubs. No where in the country has such rich assemblages of colourful flower as in temperate Himalayas where rhododendron flowers form kaleidoscopic sights miles after miles. Also myriads of species of flowering herbs and shrubs of matchless shape, texture, colour and abundance occur in alpine meadows. *Rheum nobile,* a herb in this zone produce the most beautiful flowers on earth. The flowers of *Aconite, Anemone, Iris, Begonias* etc. are of rare beauty.

There in no comparison to lotus flowers amongst the flowers of plains of India. Sweet scented champ and keora are legendary. *Ariq stolochia* flowers are swan like, petals of some flowers are bell-shaped, butterfly shaped, strap-shaped, wheel-shaped, tubular and funnel-shaped and various other types. Some of the characteristic flowering plants are jack-in-the-pulpit, bread-leaved arrow head, solomon's zig-zac, iris, asters, devils triumpets, heart of the earth, honey lotus, honey suckle, shepherd's purse, hog-apple etc. The flowers of orchids, the largest family of flowering plants in India, are exquisitely beautiful; the genera *Vanda, Dendrobium, Cyprepedium, Cymbidium, Rhynchostylis,* etc. are several such that produce flowers of commercial importance.

Floral Biology

Flowers are defined as the reproductive organ of a plant. They may be showy or inconspicuous. They may be unisexual or bisexual; male and female flowers may grow in the same plant or on different plants. Flowers may grow singly, or in a single head or in a cluster of heads, they may also grow in branches. They may be found, sometimes, on ground level growing from underground rhizomes and bulbs. They grow concealed in a fig. In arum family, grow on thick axis with distinct positions of male and female flowers.

The study of floral biology in different localities of the country has yielded valuable data and this is considered as ideal biological subjects in schools and colleges. The pleasure of a gardener is inexhaustible. The seeds of a flower contain the germ of a long series of varieties and mutations and the study of flowers may lead to several related subjects of art and science. No other occupation is so rich in potentialities.

Of all methods of pollinating agencies of flower such as water, air, insect, bird, bat, snail etc., insect and bird pollination are most complex. Various factors are involved in the process, such as colour, scent, size, length of tongue, length of mandible, length of maxilla, position of nectar etc. Bats are crepuscular or nocturnal and eat insects, flowers, petals, fruits, nectar and pollens. They have long tongue and long slender nozzle. The tip of the tongue has very fine papillae and aid in sucking nectar. They are colour blind though are able to recognise pale colour. Bird pollinated flowers are all of bright colour. Associations of bulbuls, sun-birds, honeysucklee, parakeets, barbets, pigeons, doves etc. with flowers and fruits are well known.

Pollen and Nectar

Pollen and nectar play a significant role in nature owing to symbiotic relationship with insects and birds. Pollens are of varied colours and source of protein, sugar, mineral element (potassium, calcium and phosphorus). Pollens transmit the male genetic material in the sexual reproduction of all higher plants and are well suited as a research tool for studying many patterns of plant and animal metabolism. Pollen as food of birds and insects has been widely acknowledged. The higher the evolutionary level, the more advanced is their chemical composition. Gradual evolution from amorphic to haplomorphic, actinomorphic, pleomorphic, streo-morphic and zygomorphic forms of flowers is associated with simple to complex types of chemicals in flower. Wasps do not eat pollen, but eat nectar. Bio-chemical history of pollen is complex and yet not fully explored. An analysis of Alder pollen reveal the presence of fat, protein, sugar and potassium, calcium and phosphorus.

Nectar is the product of glandular secretion of flowers. In classical mythology, nectar was considered as the 'drink of Gods.' It is a sugary secretion. Chemically, nectar contains organic acid, etherial oil and polysaccharides. Nectar forms food for very many species of insects and birds. The nectar sacks in flower are protected by hairs; the petals have various coloured patterns called the 'nectar guides' to lead the insects to the nectar. Some parts of flowers also act as nectar reservoirs. Deep seated nectar sacks are picked up by moths, butterflies, honey bee and bumble bee. There are several plant families which form large quantity of nectar. Some families form only sucrose, others equal quantity if glucose and fructose. Nectar provides carbohydrate and energy to birds. In a male flower abundance of sucrose was noticed while in female abundance of source and fructose was noticed. Various factors like atmospheric humidity, age of plants, moisture in soil etc. govern the formation of nectar.

Honey is the primary product of nectar. Honey collected from beehive in different localities vary sidely in colour, scent, taste, consistency; but nutritional value depends on the particular species of flowering plants. Apiaries in various parts of the country produce wide variety of honey depending on the type of flower available in the locality. In west Bengal apiaries produce honey from the flowers of mango, sajina, mustard and neem primarily. In Meghalaya honey from orange flower is a delicacy. Lotus honey is also considered a high value.

A remarkably high quantity of honey is produced in the estuarine tracts of Sunderbans. This is produced by the rock bee (*Apis dorsata*). Several plants contribute to the nectar of which Khalsi produce the best honey ; bulk quantity of the honey is produced from Goran, Genwa, and Hental. Man-animal, insect-animal and intra- and interanimal relationship in Sunderbans during the honey season (March to June) forms a unique field of study unknown elsewhere.

Cut Flower and Decorative Flowers

India is yet to make a headway in the growing of commercial flowers. Orchids have received some attention and some quantity is being exported. Many indigenous species have not received any attention by the plant breeders or the horticulturists. Cut flowers were extensively used in ancient Athens and Rome. In the time of Nero, the growing of flowers for the Roman luxury trade was a thriving industry. The Europians have spent fortunes on roses. Flowers add elegance to food served in ceremonial occasions. The vast floral resources of the country should be developed for commercial purposes.

Decorative use of flowers developed along with the spread of Buddhism. In India an essential part of the faith was the alter floral offering; also floral decorations were profoundly expressive of religious and philosophical attitudes. Floral decorations were used by man to embellish his environment and his person. They have played in important role in festivals, religious ceremonies, public celebrations and in courtships.

Some plants are of Gods. The use of hallucinogenic plants and flowers have been a part of human experience for many millenia. Some plants produce complex chemical compound capable of inducing visual, auditory, tactile, olfactory and gastatory hallucination or causing artificial psychosis.

These are some of the qualities enumerated on flowers. The instances cited must have testified to the essential role played by flower in the cultural life of man through the ages and how they are interwoven in the main chain of evolution, plant-animal interaction and in modulating environment in entirety. Scientists in the field of chemistry, bio-chemistry, floral biology, genetics, horticulture will find a vast field of research not fully explored as yet. It is possible to develop the floral resources of India and create a market for commercial species. Let us preserve the vast resources and develop them in a time bound programme.

A brilliant red coloured Cactus flower

A butterfly supping nectar of a flower

3

TROPICAL FORESTS – ITS BIO-ECOLOGICAL RHYTHM AND RESOURCES BONANZA

Except perhaps the tropical rain forests of Brazil, Malaysia and Zaire, Indian tropical forests have drawn the attention of the biologists and foresters for their unique geographical realms, assemblages of many floral and faunal elements and equally diverse and rich forest resources. In India, these forests are situated in three distinct biogeographical realms–one in northeastern corner dominated by Malaysian, Chinese, Japanese and Indo-Chinese elements, second in the south-western corner of the country dominated by some of north eastern elements besides African and Ethiopean elements and the third in Andaman and Nicobar Islands dominated by Malayan, Indonesian and rich endemic faunal elements, each of the areas are wide (2000 kms) apart from each other.

These tropical forests are of unlimited possibilities as they have managed to live for several million years and have undergone countless evolutionary changes. The warm and humid climate, mingled with long time span, have resulted in rich and diverse variety of plants and animals resources to be seen and believed. It is believed, and perhaps accepted, that the ancestors of human being originated in the tropical forests.

Tropical Soil

Tropical soil is poor. Because of rain, the same water that

nourishes life also takes it away. The roots are shallow, but these can absorb nutrients as soon as they are released into top soil. Millions of bacteria and microorganisms in soil aid the process. Decay of dead animals and vegetation results in accumulation of potassium, calcium, phosphorus and nitrogen. In tropical forests, nutrients are locked within its vegetal matter and are continually circulated and recycled.

The luxuriant vegetation of tropics grows on meagre soil. Like a huge bath-sponge, the land soaks up water which evaporates slowly in the atmosphere or goes to swell streams and rivers. These forests hardly draw any nutrition from the soil; nutrients form a part of a perpetual cycle between plants and animals. If the tropical forests are stripped of its wood, practically all the store of nutrients is lost. In a fire, huge amount of carbon dioxide accumulates in the atmosphere. Forest burning has contributed to 5 percent increase in carbon dioxide stops the escape of head from the earth. Half of the rainfall of the Amazon basis is contributed by the tropical forests. These are some of the qualities of tropical forests.

Resources

Tropical forests have specific pattern in their distribution of floral and faunal species. Myriads of species are found. It is said that tropical African forests have 25,000 plant species and Malaysian forests about the same, but Brazilian forests have about 50,000 plant species. Biologists think that there is a connection between the tropical forest diversity and its long survival. Indian forests are not as rich as the aforesaid forests, but have more than 18,000 species.

Trees in tropical forests are of varied height, the mighty ones reaching more than 50 metres height and the forests are multistoreyed. They are top heavy and strong wind topple them. Some trees develop buttress and silt roots. Lianas (thick woody climbers) are often 50 metres long. The forests are evergreen and have abundant epiphytes, mosses, lichens, ferns and vascular epiphytes. These epiphytes have their aerial roots which absorb moisture from air. The nutrients are obtained from leaf mould and debris that collect in the crevices of branches. The soil feeds the epiphytes as well as shelter beetles, earthworm, ant and other insects. Scores of instances can be sited about special floral adaptation, flower colouration, pollen and nectar production in a tropical forests.

There are innumerable patterns of petals, position of flowers or inflorescence in a plant to facilitate pollination. In Central and South America, more than 2,000 species of orchids depend on a family of jewel-coloured bees for pollination. The magnificent natural feature is that a variety of perfumes are produced by orchids which attract the male Euglossine bees–a family related to honey bees and bumble bees.

Tropical rain-forests have one of the most complex ecosystem on earth. Once the forests are clear felled, the poor soil does not support agricultural crop for more than two to three years. Once the vegetation is stripped off, barren soil only remains, yet much is not known about the complex ecosystem. Principal character of the vegetation is ever-green and seasonal changes only affect individual species rather the whole forests, which consists of a luxuriant mixtures of very diverse species forming such strait of vegetation. The trees are interlaced by long woody lianas and climbers and many epiphytes. There are little herbaceous and shrubby vegetation on the floor. The multistoreyed forests can be divided into several main layers, each having a typical biological character. Birds of prey nests in topmost emergent layer. The next layer below is the most important and occupied by monkeys, squirrels and brightly coloured birds. The next layer is frequented by leopards, bears, tree frogs, langurs and many other animals.

In tropical forests, there is no synchronised rhythm of spring or autumn. Some trees produce bloom and fruit twice or thrice a year, and some flowers and fruit throughout the year. In fact, the phenology of tropical forests is yet to be fully known. In the dark and damp floor, the smallest plants live in an eternal twilight while on tree-top strong light and wind sweeps the perching birds. The immense variety of plants in tropical rain-forests offer a still unexplored treasure trove of species for cultivation and use in medicine. From wild species mankind have been able to grow coffee, tea, cocoa, banana, yams, palm oil, papaya and very many species of food crops. These forests have been found to be of great importance as genetic reservoir of cultivated plants.

Animal Life

Tropical rain-forests are the areas of evolutionary history. The animals are extremely sedentary. The assemblage of rich number of species within a small area is a special feature of such forests. In South-east Asia, there are 220 species of mammals. In one hectare (ha), as many as 300 to 400 species of birds are sighted. In Africa, 40 per cent of the mammals of rain forests live on trees. Most of them are solitary. Almost two-third of the mammals are nocturnal living in hollow of trees, on branches, on leaves or on foliage of epiphytic plants. Such forests in India are equally rich, but thorough investigation is yet to be made.

Tropical forests are multistoreyed where ground layer, herb layer, shrub layer, low-tree layer middle-tree layer and top-tree layer are prominent. Myriads of leaving creature such as leaf-cutting-ant, fungi, bacteria, mites insects that decompose dead leaves live in the ground layer. Herb and shrub layers gather moisture and many species of insects collect and provide ground dwellers food and shelter. Mosses and lichens grow on the crevices of the tree-trunk where beetles, caterpillars and moths take shelter. Middle and low tree layers trap heat and moisture. Middle-storey trees harbour most of the arboreal birds and animals. Monkeys, langurs, squirrels, lizards, snakes, toads and mammal also inhabit tall evergreen forest trees and rarely come to the ground. There are many flying squirrel and gliding creatures, scaly-tailed flying squirrel and gliding creatures, scaly-tailed flying squirrel, tree pangolin that eat termites, ants etc. From tree canopy to ground floor a tropical forest, it is clear, provides millions of habitats where animals make their home.

Warm and humid tropical forest climate presents an ideal habitat for cold-blooded animals to survive. These animals occupy every layer of forests from ground layer to the top and have remarkable power to masterfully camouflage and to mimic. Large as well as small snakes are residents of tropical forests. Many of them are tree-dwellers and some can glide from tree to tree. Millions of frogs inhabit trees, some mimic leaves, some glides, some crawl, some are multi-coloured and few are poisonous. Some are nocturnal and some flying. The tree frogs lay eggs on water that accumulates among the rosette leaves of epiphytes. Scores of similar examples can be cited.

Predator-prey relationship in a tropical forests is a unique subject of study. Graminivores feed on buds, leaves, flowers and fruits, predators are carnivores who feed on flesh. The predators are consumers of the second to even sixth level or so, or even higher.

Royal Bengal Tiger

Elephant recently captured in mela shikar

Role of Ants

The role played by ants in interaction of various parameters of environment is not fully known. Owing to intense competition for food and space in tropical forests, interactions between same species of plants and animals, between plants, animals and environment are more subtle and intricate than in other forests. Ants are perhaps the only single group of creature that change environment to a considerable extent. Ants are the largest group of insects in the tropics. In tree barks, roots, branches, leaves, fruits and flowers they are everywhere. There are several species and are coloured red, black, brown and of other mixed colours. Leaf-cutting insects are of immense importance and they live in soil. They cultivate fungus to feed the young. The ants strip off leaves from the vicinity of their nests. They ingeniously solve the problem of garbage disposal (they have underground garbage disposal areas). From ant to fungus to ant and back to tree root, the nutrients of the forests are used and recycled. This is an example of the efficient use of energy that helps to make tropical forest a stable ecosystem. Ants are carnivores and feed on arthopods (spider, grass-hoppers, crickets, cockroaches, etc.). More than 200 species of scavenger flies depend on army ants for their food supply.

Bat-forest Interactions

Bats in tropical forests are nature's gift. Nocturnal as they are nothing much is known about the vital role played by them in tropical forests ecology. They are one of the most diverse group of animals in tropical forests. Some of the bats are insectivores some frugivores, some nectrivores, some carnivores and some vampires. How these diverse feeding habit evolved in bats is a complex study. They have tremendous capacity to co-exist at a very short space. To locate food in the dark, most bats use non-visual techniques. They use a kind of ecolocation or sonars to find their way. As they fly through the dark forest, they continually send out a series of high-pitched clicks at frequencies far beyond the range of human hearings. The sound waves from the clicks bounch off trees, blossoms, insects whatever may be their pathway.

Bats pollinate flowers at night. They prefer white drab white in colour or greenish or brownish; the flowers are large and have unpleasant odour. It is believed that vegetation seem to have evolved along with fruit-eating bats. Vampire bats are blood suckers. Carnivore bats forage at forest flower for their food. There are fishing bats too. Like birds, bats also live at all levels of forests.

Scientists think that bats have sufficient impact on the food wave, because they are the most massive source of protein in the whole of tropical forest system. Total impact of bats on environment has to be worked.

Tropical Rain Forests of India

Indian rain forests are dominated by *Dipterocarps.* There are only nine species of *Dipterocarps* besides many other associate tree species. In Malaysian region, there may be 60–250 species of the members of *Dipterocarpaceae*. The rain forests of Assam is dominated by two species such as *Dipterocarpus macrocarpus* (Hollong) and *Shorea assamica* (makai). In south-western region (at Makut, Nilumbur, Coorg, North Canara, Palghat, Wynaad, Tiruneveli and several other places), *Dipterocarpus* has *Palaquium, Mesua, Cullenia, Poeciloneuron, Calophyllum* as principal associates. In Andaman, there are four species of *Diptercarps* with associates besides *Mesua, Terminalia, Pterocarpus, Artocarpus, Calophyllum, Hopea* etc. Only 40 species, out of total 200 tree species, are being commercially exploited. Tropical forests of India are primarily providing plywood, match-wood and packing-case wood besides yielding some quantity of constructional and furniture timber. Production per hectare is about 300 m^3 which is highest in any forest type in India. Large areas of tropical forests in Nagaland, Cachar hills, Assam, Mizoram, Manipur and in Arunachal Pradesh have been felled in the practice of shifting cultivation. Forestry operation in achieving natural regeneration by felling on non-commercial species has also changed the nature of forests. Mixed forests have been converted into pure forests of a few selected species. Intense biotic activities have completely changed the nature of forests in both north-eastern and south-western regions. At present, only selected areas remain as preservation plots and biosphere reserves to represent typical tropical rain forests. Wildlife sanctuaries and national parks are also indirectly protecting some isolated spots.

Tropical rain forests of Andaman and Nicobar Islands are pride of the country. It extends over about 6,500 sq km and still have substantial areas untouched forests. High degree of endemic flora and fauna is an asset of the tract. Besides, it has several species of animals not found anywhere in the world. Mention may be made of Narcondom hornbill, Nicobar megapode, Andaman teal, Crab-eating macaque and Robber crab. It is said that about 35 per cent of the flora is endemic and, therefore, the forests need complete protection. There are 250 species of avifaunal species, besides many species of invertebrates with substantial number of endemic species.

Destructive forces have also affected these forests. Refuges rehabilitation (25,000 East Bengal refugees) has been responsible for destruction and clear felling of about 1,000 ha. Several areas have been completely depleted. Selection felling done to ensure natural regeneration has opened up extensive areas of forests which have lost the character of tropical forests. Many areas have been exposed to erosion. Most significant change and depredation has been caused by severe cyclone which is a regular feature of Andaman forests. Hundreds of square kilometres of forests have been uprooted and damaged (shallow root system and heavy crown of trees).

Most disastrous human activities have been the conversion of virgin tropical forests of Little Andaman into oil-palm plantation. This species (*Elaeis guinnensis,* an exotic) has been planted over 2,000 ha by felling of forests. It has been planned to plant up 50,000 ha of the total area of 73,400 ha of forests. Areas planted up with oil palm have shown surface erosion and no plants have come up under the shade of this crop. The population of the islands now exceeds 130,000 and has perhaps reached the saturation point. This has to be discontinued as the water-retention capacity of the spoil is poor, there is dearth of source of drinking water, many cyclone-prone areas which force the necessity of preserving the last vestige of tropical forests.

There is no parallel to the richness of faunal elements in tropical forests of India. Mention may be made of tiger, leopard, elephant, various species of deer, sambar, gaur, besides many other species in the forests. There are innumerable species of insects. Most interesting character of the forests is the occurrence of several species of primates such as the hoolock, the bonnet macaque, the rhesus monkey, the Assamese macaque, the lion-tailed macaque, the common langur, the capped langur, the Nilgiri langur, the slop lorris and slender loris.

In this short presentation, the glory and destruction of this invaluable gift of nature have been sung. Imperative necessity of absolute conservation of forests has been stressed. Public consciousness has been responsible in saving the silent Valley forests from complete annihilation. Voice should now be raised to protect the virgin forests of Andamans and retain their pristine glory. The tropical forests of mainland have been depleted almost completely. The last vestige of these forests have now to be protected.

Should human being destroy forests that he can never replace ? Should the forests be destroyed where many myriads of creatures evolved ? Should they lose the oxygen produced by these forests ? Would there be any substitute for these forests that provide vast quantity of various kinds of timber ? Should they convert the fertile soil into a mass of sterile soil and invite desert into a sylvan tract ?

These are a few questions amongst many which the foresters and administrators will have to answer before more forests are clearfelled or converted for other purposes ?

4

GLORY THAT WAS INDIA'S VIRGIN FORESTS

Will it be too much to say that India has only vestigal remains of her once glorious virgin forests ? This writer initiated a research project to delineate the amount of virgin forests in India. The findings revealed that only about 10,000 sq km of forests could only come close to the category of virgin forests. In fact few forests are now really virgin, retaining their pristine glory.

Several Government Departments were interested in the above research project. It was felt that there was an imperative necessity to delineate such complex ecosystem as in tropical forests and manage them as biosphere reserves; for these virgin forests shall meet to the need of present and future human population and original character of the biotic communities within the natural ecosystem (safeguarding the genetic diversity on which their evolution depends), provision of ecological and biological research are some fraction of qualities for which virgin forests are to be rationally managed. Management is also necessary for all ecological, genetic, behavioural, evolutionary and physical processes in such natural forests.

If the findings of this writer's research project (that India is left with only 10, 000 sq km of near virgin forests) were correct (there is no reason to believe that they are not correct), then what had happened to our grandiose claim that India has vast tracts of virgin forests ?

In the September 10, 1984 issue of The Hindustan Times, in his article entitled "Indian Forest Policy in Doldrums", this writer had given his research findings about the total vegetation cover of the country which comes to mere 20 per cent of which 7 per cent was in somewhat good cover, while the balance 13 per cent was open and depleted. In that article, the writer tried to bring into public focus how the present management practice of Government and the pressure of human and cattle population accentuated by socio-economic condition led to the present disastrous situation.

In the present project of delineation of virgin forests, all the Chief Conservators of Forest of the States and Union territories ,were consulted to collect information and compile the report. The response of the Chief of the forests was, by and large, spontaneous and overwhelming. They all agreed that there were no forests in their states till in the pristine virgin glory. Actually, barring a handful of forests situated in inaccessible high altitude and remote places, there is hardly any area which has not been depleted because of human interference, grazing and fire.

To trace the reasons for this deplorable situation, it is necessary to have a look at the physiographic past of the country as recorded in the old scriptures and in other works. During the Ramayana and Mahabharata eras, the forests were considered to be of great culture and aesthetic value. Forests and trees were treated with great reverence. Almost entire country was full of forests. The Chinese pilgrim Huen-Tsang in his travel account mentioned about extensive and dense forests in the country. We also find in the recorded history of Alexander's invasion, the reference of dense forests in north-west India which is now a desert. During the medieval period, forests suffered heavy depletion due to foreign invasion and foreign rule. that situation was partially restored during the British rule when vast areas of forests were brought under reservation and forest law was enacted. But vast areas of forests were felled to raise tea, coffee and rubber plantations. Rights and privileges of the people were acknowledged in the forests.

Within the broad climatic zones (tropical, temperate and alpine) sixteen different climatic climax types have been identified in India. More than 6.4 million hectares (ha) are under tropical rain and semi-evergreen forests. Another 22.4 million ha are under tropical moist forests. But unfortunately the density of these forests is now poor, productivity low and stock per ha is far too low in comparison to similar forests of other countries.

In the last two paragraphs, the lost glory of India's forests have been adequately highlighted. But some forests were still left in near virgin state. But why such forests need protection? What are their qualities ? The trees of virgin tropical rain-forests are tall, covered with moss, lichen and climbing plants. These forests have been growing for centuries. Generations of animals have made their home here. Ferns and orchids have created vertical ecological biotopes in the forks of branches and in laminations of leaf bases. Lizards and squirrels have taken permanent shelter in the branches and the butterflies flutter in clouds around the highest tree tops. The leaves are varied, variously coloured which change very often. Some trees bloom and bear flower and fruit two to three times a year. Immense variety of plants offer still unexplored myriads of species. In the jungles grow wild plants from which are obtained coffee, tea, palms, yams, cocoa, banana, etc. These forests are also important as a genetic reservoir of very many cultivated plants. Such forests are arenas of evolutionary history. here, stranglers encircle it's host tree, the tree fungi help the process of decomposition, lianas form impenetrable tangle offering almost impregnable passage even for wandering wild elephants.

The Chief Conservators of Forests must have kept all these qualities of virgin forests in their full view and initially could not identify any virgin forests. Later, on second thought, identified a number of wildlife sanctuaries and national parks and other small patches of forests which were near virgin and somewhat retained their original characters. A state-wise list of such forests were prepared as under :

Andhra Pradesh : Fifty one small and isolated sites have been identified which can hardly be given protection. These sites may not serve the requirement of biosphere reserves. Four reserves viz. Nugur, Singaram, Tadicherla and Tavadi each of which covering more than 100 sq km have been identified beside other seven each with an area of just over 50 sq km. About five wildlife sanctuaries still retain good forests.

Assam : No virgin forests have been identified. However, Bamba, Dholpani, Chanduar, Nomeri, Borail and Subansiri reserves are being considered for permanent protection as virgin forests. A few wildlife sanctuaries have been mentioned which retain good forests.

Bihar : No virgin forests have been identified.

Gujarat : Apart from six wildlife sanctuaries and national parks, only one area viz. the core area of Jammwalagir lion range park covering 140 sq km has been marked as virgin forests.

Himachal Pradesh : There are no areas to mark as virgin forests, though a few wildlife sanctuaries are in good shape.

Jammu and Kashmir : Other than a few wildlife sanctuaries, seven high-altitude national parks and bioshpere reserves covering more than 100 sq km have been marked as virgin forest.

Karnataka : Dandeli wildlife sanctuary has a very large area in a good state. About 1,000 sq km of reserves in the forest blocks of Jog, Attigod, Govardhangiri, Nagarhole, Bandipur and Mahadeswara have been marked as virgin forest.

Kerala : Beside seven wildlife sanctuaries and national parks, well known reserves at Silent Valley, Sultan battery, Periyar and Parambikulam have been marked as virgin forests.

Madhya Pradesh : This state, having largest forest area of the country, has hardly any virgin forests. Wildlife sanctuaries national parks covering about 1,200 sq km have been marked as virgin forests.

Maharashtra : Beside a few wildlife sanctuaries and national parks, there is hardly any area to mark as virgin forests.

Meghalaya : About twenty scattered areas have been marked as virgin forests of which Saipung, Narphu and Mongkyallem covering a total area of 436 sq km are in very good shape. There are small areas in Bagmara, Rongrenggiri, Angra Toli and Chimabongshi reserves of about 20 sq km each which are virgin.

Mizoram : Some good forests in Dampa and Chamdur reserves have been marked which cover an area of 300 sq km.

Nagaland : Intanki reserve, Dzubhu valley reserve, Central, Eastern and South-eastern reserves covering about 781 sq km have virgin forests.

Orissa : Other than Simplipal, Satkosia and Bhaittar Kanika reserves, there is no other forests to mark as virgin forests.

Rajasthan : There is no area to make as virgin forests.

Sikkim : Sixty one reserves have been marked as virgin forests. Except Kanchanjangha National Park, there is hardly any large area. Virgin forests may be marked in the reserve of Latu, Pangola, Tolung, Lik, Zongri and Lapdang.

Tamil Nadu : Beside some wildlife sanctuaries, thirty seven areas of virgin forests have been marked of which Madumalai, Kalakadu, Singampatti and Verrapuli covering about 760 sq km may be mentioned.

Tripura : Thirteen isolated forests have been identified of which notables are Samruhalai, Baramura, Deotamura and Atharamura covering about 459 sq km may be mentioned.

Uttar Pradesh : No virgin forest have been marked.

West Bengal : Neora valley forests have only the growth potential and qualities of virgin forests.

It is worthwhile to mention here that Arunachal Pradesh and the Andaman & Nicobar Islands have substantial areas to mark as virgin forests. The Jarwa reserve covering about 700 sq km beside some forests in several islands in the Andaman should be marked as virgin forests. Present tinkering with various areas must be stopped. Clearfelling of tropical forest in little Andaman for oil-palm plantation must be halted forthwith. Namdhapa, sunderbans and other reserves marked as biosphere reserves and wildlife sanctuaries in Arunachal pradesh, West Bengal and elsewhere are wise move to preserve the virgin forests.

Let there be at least two states in the country where virgin forests in true sense of the term will remain. Let the forests of Andaman and Arunachal remain untouched from human vandalism and management in the name of economic amelioration stopped.

5

ENDANGERED AQUATIC HABITAT

While the orbiting Satellites have contributed greatly to our knowledge of cloud cover, cloud-top-temperature, vertical temperature, humidity level at different altitudes and Remote Sensing Satellite on Ocean temperature and geostationary Satellite on weather developments, it is yet not known if any of these Satellites have reflected data on moisture level of various types of soils and the biologists have correlated such finding with occurrence of micro flora and fauna in this country.

The Department of Space had recently (1983) staged a symposium at Ahmedabad and some years back the National Science Academy held a workshop on hydrology. It is not known either if any study has been done on floral and faunal assemblages on their biomass vis-a-vis moisture regime.

The term aquatic habitat covers inland water, ground water, fresh water such as groups of brooks, rivers, tanks, ponds and swamps. Estuarine swamps are included in the term, besides cracks in rocks, space in rocks, gravel beds which accumulate water and moisture, Fresh water that forms a fraction of total water on earth, has therefore, posed a problem in this expanding population and demand for more space for habitation and cultivation. Nevertheless such areas are extremely varied during monsoon months and identification and mapping aquatic surface pose a problem.

International Union for Conservation of Nature and Natural Resources (IUCN) offers a broad definition of wetland as "Areas of submerged or water stagnant lands whether both natural of artificial or permanent or temporary of whether the water is static or flowing or fresh, brackish or salt water dominated areas to be considered would include–marshes, sloughs, bogs, swamps, fens, peatlands, estuaries, bay, lagoons, ponds, lakes, rivers, springs and reservoirs; where marine or coastal waters are involved water upto a depth of 6 metres at low tide are included".

The present requirement is far too extensive and much too complicated. The author has examined Landsat imageries of band 5 Black and White to identify such land but not much reliable data was available. He also examined a number of scenes on 1 : 1 million scale of colour composites on band 4, 5 and 7, but the findings could not help identification of wetlands and swamps. He considers that NOAA Satellite pictures could be useful for such work but it could not be analysed.

It has been possible to identify large dam sites, reservoirs, aquatic wildlife sanctuaries, mangrove forests, back waters lagoon. But what about innumerable smaller sites and reservoirs which are created during monsoon ? Such small sites of water and moisture laidden pockets are scattered all over the country. What also about the moist soil all over the country that harbour millions of invertebrate fauna and identifying areas of rare plants, endangered plants, concentrated zones of Myxophyceae, Chlorophyceae, Bacillariophyceae, various soil, inhabiting microbial flora and fauna; in estuarine region the concentration of molluscs, crustaea, rich intertidal floral and faunal zone, rich zone of *Forminifera, Noctiluca* and *Dinolflagellates* ? What was about identifying concentrated zones of *Platanista, Crocodylus porosus, Panthera tigris, Axis axix* in Sunderbans ?

The estuarine fauna are diverse and ore of mixed types. There are multi-dimensional zonations by animals that make any study much too complicated.

Hydrological Data and National Picture

The hydrological regime of the rivers is governed by the precipitation of the monsoon primarily. Local climate, topography, soil status also governs the hydrological regime. Human intervention complicates the situation.

The situation in the Himalayas, eastern and western ghats, Vindhya and Satpura hills testify to disastrous hydrological situation in the country. Bulk of the country suffers from drought and flood every year. Water holding capacity of soil has been reduced to minimum. But water is the basic resource for irrigation, hydroelectricity and for daily uses. Constructions of dams have complicated the issue. Even in such a precarious situation neither hydrological survey has been made nor aquatic environment has been identified and so far all these have been the subject of imphirical investigation.

Water Surface

An FAO expert on Inland Fisheries in South-Eastern Asia was asked by the author about the extend of inland water surface India. He observed "Some one long back worked out a figure and that was now being quoted; there was no authentic basis of data often quoted by Indian Govt.".

According to the report of Indo-Pacific Fisheries Council (1952) inventory of aquatic surface (km^2) was made for the Country as follows :

(i)	Fresh water pond	-	3,379	km^2
(ii)	Brackish water pond	-	26.7	km^2
(iii)	Natural Lakes	-	1,800	km^2
(iv)	Artificial Lakes	-	7,258	km^2
(v)	Rice field	-	57,644.2	km^2
(vi)	Estuarine Lagoons	-	2,644	km^2

This observation does not satisfy the author's need. Taking it for granted that these figures have some basis, the following are yet wanting :

(i) Aquatic sites in glaciers and in Alpine situation

(ii) Aquatic sites among rocks and boulders

(iii) Aquatic sites in crevices of rocks

(iv) Temporary aquatic sites during monsoon period

(v) Moisture regime in various types of soil and status of underground water at various seasons.

The figure quoted by I.P.F.C. seem to have some lacunae, as they quote Thailand and Indonesia which have 61,868 km^2 and 1,35,480 km^2 of water surface respectively; India is many times larger than Thailand and have just little more water surface.

A survey carried by the Forest Department, West Bengal shows that the State (within Reserve and Protected Forests) has 57,644 ha. (576.44 km^2) of Wetland. In 24-Parganas District (Outside R.F.) the Land and Land Revenue Deptt. have identified 1050 ha of marshes and bills which seems to be very poor. There are many bills outside Reserved in protected forests in the districts of Murshidabad, Malda and West Dinajpur.

Mapping of all smaller sites is difficult.

National Parks of Wild Life Sanctuaries in Marshy and Aquatic Situation

At least twenty five (25) Wildlife Sanctuaries and National parks have been established in India on such sites. Notable among these are Lake Pulicut (actually it is a lagoon), Lake Chilka (it is a lagoon), Point Calimere, Vendanthangal, Villangudi and Ghana. Others are Nealampalli Nalsarovar, Sultanpur, Gobind Sagar, Ghata prova, Rangan thitoo, Sirpur, Gandi nagar, Keibul lamjao, Bhitar kanika, etc. Several other sanctuaries have ponds and dam sites or river inside; some such areas are Corbett National Park, Sibpur National Park, Taroba National Park, Ranthambhore Tiger Reserve and several others. The largest water area, however, has in Mangrove forests of West Bengal (about 4,500 km^2).

Aquatic and Marsh Plant Resources of India

This country has about one hundred eighty (180) *species of aquatic plants (Marshes and damp site have very many other species); These belong to ninety (90) genera and thirty nine (39) families. There are 9 species of Pteridophytes, 75 species of dicot and 100 species of monocot and vascular plants. The aquatic plants are either floating, or submerged or amphibian or floating with roots at tank bottom. Some characteristic features are :

(*i*) Widely distributed species number 35, less widely distributed number 38 and restricted number 12. Temperate saline number 24 and restricted species to Indian Sub-continent number 35.

(*ii*) Quantitative abundance of species are *Eichornia crassipes, Leersia hexandra, Trapa bispinosa, Pistia strateotes, Paspalidium germinatum, Scripus* Sp., *Ottelia alismoides.*

(*iii*) Temperate saline species are *Aldrovanda vesiculosa, Callitriche stragnalis, Halophila ovalis, Hydrolythrum wallichii, Ruppia maritima, Utricularia inflexa* and *Zannichellia palustris.*

(*iv*) Some species restricted to this country are, *Aponogeton echinatus, Limnophila aquatica, L. connata, Myriophyllum tetrandrum, Najas brevistyla* and *Woolfia mircoscopica.*

(*v*) Heavily exploited species, which form food stuff, are threatened may be endangered soon are – *Nymphaaea nouchali, N. stellate, N. alba, N. tetragona, Nelumbo nucifera, Trapa bispinosa, Endydra fluctuans, Ipomoea aquatica.*

(*vi*) Biologically important species are *Lemna* sp., *Azolla* sp., and *Spirodela* sp. which are avidly eaten by fish as well as bird population. Fruits of most of the species are eaten by aquatic birds, besides leaves. *Azolla* fixes nitrogen.

Utricularia and *Aldrovanda* species which do not occur widely should be protected because of their biological importance and limited distribution.

Najas graminea, a monocot that occurs densely in the swamps of Ranthambhore Tiger Reserve, forms a staple food for Sambars during dry months and needs thorough examination for nutritional value besides its protection.

(*vii*) There is abundance of occurrence of algal flora in aquatic situation which is also a major component of planktonic flora. As benthic or bottom flora Diatoms are extremely important.

Criteria for Identification of Wetland

Ramsar Convention (convention of Wetland of International Importance) has fixed some quantitative criteria for identifying wetland of importance to (*i*) water fowl, (*ii*) to plants and animals for (*iii*) assessing the value of representative or unique wetland. A weltland should be considered of international importance if it (*a*) supports the appreciable number of rare, valuable and endangered species of plans and animals; (*b*) harbours endemic plant species; (*c*) maintains the genetic and ecological diversity of a region because of the quality and peculiarities of flora and fauna; *(d)* has plants and animals at a initial stage of their biological cycle.

It is necessary to evaluate the wetland and marshes according to the criteria fixed by Ramsar Convention on wetlands of International Importance especially as water fowl Habitat signed in 1971. For water flow of international importance, a wetland should have 10,000 ducks, geese and swan; or 10,000 coots; or 20,000 waders and these should support 1 per cent of individuals in a population of 1 species or 1 per cent of the breeding pairs in a population of one species of water fowl. These criteria put only a few areas under wetland.

Aquatic Avifauna vis-a-vis Sites

Diversity is the essence of study of aquatic birds. The story of aquatic bird is the story of water drainage and siltation. Man made ponds have only by-product value of such birds. These wetland sites are a living, dynamic entity and never static, always changing plankton and other vegetation have high nutritive value.

More than one hundred aquatic birds, both local and migrants, visit the aquatic sites all over the country. Specific diversity is high in very many species. Many birds in their breeding season burst into activities to collect food material for their fledglings. Hundreds and thousands of species of vertebrates and invertebrate animals are collected by the breeding birds to feed their fledglings. Various predators prey upon the fledglings and the natural selection made in the battle of survival of the fittest.

Myriads of gramnivorous avifauna need wide aquatic surface to wade and swim and they eat leaves, flowers, fruits, stems and roots of various aquatic weeds besides phytoplankton and zooplankton.

The requirement of some of the species of birds are specific. It is needless to mention various species of waterhen, crake, coot, lapwing, sand piper, shank, gull, tern, cormorant, darter, ibis, heron, egret, teal, goose, mallard, shoveller, pintail, white eye, grebe, crane, curlew, plover, jacana, slilt, snipe, skimmer, openbill, stork, spoonbill, paddybird, bittern, etc. and some rare and prized birds like flamingo and pelican are integral part of wetland.

Aquatic Communities in Sub-alpine and Alpine Hills

At high altitude there are insect associations on stagnant water lake, on fresh-water pools, on salt and alkaline water lake, on torrential streams and on hot springs. There are dearth of surface hunting species of the order Heteroptera and Coleoptera. Some high-altitude pools are of glacial origin. In torrents there are few species, but abundance of individuals. On the shores of lakes extensive association Collembola-Staphylinidae occurs. In hot sulphur spring few species occur (Diptera predominant.)

Communities in torrential streams are (I) *Ephimerida - Plecoptera* association, (II) *Denterophlebiidae-Blepharoceridae-Simuliidae* association, (III) *Chironemedae* consociation, and (IV) *Trichoptera* consociation. Nos. (I) and (II) are submerged stone association. No. (I) association has five prominent genera, viz. *Bactistron, Eidyurus, Capnia, Nemoura* and *Rhabdiopleryn.*

There are specific insect life on damp lichen and moss covered rocks, wet standing boulders, snow surface, under snow, edges of snow and in glacier creavasses.

Aquatic Insects

There are a variety and abundance of insect life in water. They slide and twirl across the silvery expanse of water. The abundance of damsel flies, may flies, caddis flies, dragon flies, and crane flies which form aquatic communities fall prey to the fish, toad and frog and avifauna population. Larvae of aquatic insects can thrive in hot spring, cold-glacial streams and on various other types of waters. Other common insects of ponds and brooks are the back swimmer, the water boatman, the pond skater and the whirligig beetle.

The drying up of a pond is a disaster to aquatic life. Some species can, however, endure by burying themselves in mud bottom and await return of water. Some diving bettles and water bugs are powerful flier and leave drought area when necessary. The importance of aquatic insects that form primary food of aquatic ducklings and insectivorous birds need not be stressed. These aquatic sites need identification and protection.

Aquatic Animals

The Gharial (*Gavialis gangeticus*), water monitor (*Varanas* sp.,) the Mugger (*Crocodylus palustris*), Estuarine crocodyle (*C. porosus),* Gangetic dolphine (*Plantanista gangetica),* Green turtle (*Chelonia mydas), Hawbill turtle (Eretmochelys imbricata), Laggerhead turtle*

(Caretta caretta) and Leathery turtle (*Dermochelys coriacia)* are already in the list of endangered species.

Species of turtle and tortoises (*Geomyda, Kachua, Chitra, Testuda* and *Lessemys* sp.) are fighting for their survival in the North Bengal districts and elsewhere for shrinkage of aquatic habitat. Very many species of invertebrates may be noted.

Algae of Inland Waters

The important of algae is diverse in aquatic environment. Myxophyceae play a key role in the hydrobiology of inland water. Some fix atmospheric Nitrogen and can be used as fertilizers in rice field. They can help to purify sewage and polluted waters. Diatoms grow on water poor in nitrates and rich in silica. Chlorophyceae forms food for many fish and invertebrates. Filamentous Zygophyceae such as *Spirogyra* forms floating net and pollutes water as they rot; they also contribute to biological purification in shallow water. A study of *Salvinia auriculata* in Cochin back waters has shown that the following groups of organisms are associated with it – *Nematodes, Annelids, Copepods, Amphipods, Isopods, Ostracods, Deiapods, Insect larvae, Bivalves and Gastropods.*

Areawise distribution of mangrove and estuarine forests are as under : Sundarban mangrove – 4,18,888 ha, Mahanadi delta (Devi and Mahanadi estuarine – 12,000 ha, Krishna delta – 5120 ha, Godavari delta – 13,304 ha, Gulf of Kutch mangrove forests – 10,000 ha, Salsitte Island and Western Coast mangrove forests – 62,208 ha, Kankinada back waters – 33,000 ha, Western Coastal areas – 52,616 ha. As such barring Sundarbans, the mangrove and estuarine forests extend over a very limited area and are very much disturbed. Bulk of these forests have been exploited and felled. Nevertheless, extensive research work has been done on the fauna of sandy shore, clayey shore, beaches and of intertidal benthic and pelagic zones. Pollutions have affected some such coastal and lagoon areas. The forests having no width and being of small patches are very much vulnerable to human depredation.

Phytoplankton of Hooghly-Malta Estuary

Blue-green algae (*Myxophyta*) distinctly predominate in Hooghly estuary while green algae (*Chlorophyta*) predominate at upper zones.

Most important components are diatoms. At least 50 species of diatoms are common (*Synedra, Coscinodiscus, Surirella, Melosira, Navicula, Diatoma, Lauderia, Biddulphia, Bacillaria* are a few genera). Thirty species of Chlorophyta are common (some of the genera are *Pediastrum, Closterium, Cosmarium, Spirogyra, Eudorina, Chlorella, Zygnema, Ulothrix, Volvox, Oedogonium* etc.). Of the 20 species of *Myxophyceae Microcystis, Phormidium, Nostoc, Oscillatoria, Anabaena, Lyngbya* etc. are common. There are seasonal variations in the abundance of occurrence of various algae.

Zooplankton of Hooghly-Malta Estuary

Copepods constitute the bulk zooplankton even though rotifers and larval forms out number them. *Rotifers, Protozoans* and *Cladocerans* are found in abundance in less saline areas. The pattern of fluctuations of zooplankton runs parallel with phytoplankton.

Common genera of flagellates are *Euglena, Dynobryon, Ceratium, Phacus, Peridinium, Noctiluca* and *Trachelomonas.* Of the protozoa common genera are *Dijtugia, Bursaria, Vorticella,* etc. *Brachionus, Keratella, Trichocerca, Filinia, Hexartha, Notholca* and *Ploesoma* are common genera. Of the copepods, common genera are *Diaptomus, Cyclops, Paracalamus, Microsetella,* etc. *Bosmina, Bosminopsis, Daphnia, Moina* and *Ceriodaphnia* are common genera.

Fish Population of Hooghly-Matla Estuary

More than 150 species of fish occur in this region. These are more than 150 bheris in reclaimed and embanked land (3500 km^2). The fish seeds are mainly of mullets, perches, threadfins, catfishes, prawns, carps etc. Fish production in India in estuarine in about 7000 M.T. annually.

Maximum catch is formed of prawn species viz. Bagda Chingri (*Panaeus* sp.), Chapra Chingri (*Panaeus* sp.), Honye Chingri (*Metapenaeus* sp.), Chamme Chingri (*Metapenaeus* sp.), Golda Chingri (*Macrobrachium* sp.) Goda Chingri (*Palaemon* sp.) and Ghona Chingri (*Leander* sp.). Other species which are common, consist of Bhetki (*Lates* sp.), Parse (*Mugil* sp.) Bhangon (*Mugil* sp.), Nona Tengra (*Mystas* sp.), Gurjaoli (*Elentheronema* sp.) Chanda (*Scalaophagas* sp.) etc. Hilsa ilisha is pretty common in the estuary besides Clupeids, Bombay duck (*Harpodon* sp.) etc. Ribbon fish (*Trichiurus* sp.).

Plant-animal Relationship

There are specific interrelation between some plants and molluscs and cruststacea. For instance, *Assiminea brevicula* prefers to live on trees, on leaves and on branches, *Neritina depressa* prefers trunks and lower branches, *Martesia striata,* Banksia edmondsoni and *Teredo juttingae* prefer roots, trunks and submerged branches, Generally *Littorina* and *Neritina* are surface burrower. Above tidal level occur *Telescopium Cerithidea, Pythia, Eilobium* etc. Innumberable such examples can be cited on estuarine and mangrove tracts. Of the plant *Sonneretia apetala, S. acida, Avicennia* sp. and *Xylocarpus* sp. should be protected from felling.

Very little is known about estuarine mammals and the ecology, but the effect of mud or sediments in water is very much perceptible. Some of them are rare. Their movement is guided by availability of food. Flesh of *Platanista gangetica* sp. is edible.

Problems

Calamities to the wetland sites have been brought by miscalculated and ignorant human activities. The swamps have been reclaimed and brought under agricultural crop; various sites have been poisoned by retting of jute fibres, use of pesticides for eradication of mosquitoes and for drainage.

Indiscriminate shooting of birds have been done by hunters all over the country which had diluted the population of game birds. Many water-birds have been regularly trapped in the villages especially during breeding season, besides illicit collection of thousands of eggs and fledgling to provide delicacies in the restaurants of the cities.

Unique aquatic sites like one in Tale valley in temperate hills of Arunachal Pradesh (near Zero) which are like Sphagnum bog of the West have been converted into cultivated land. Cattle grazing in the swamps have diluted the species and their individuals.

Plant and animals succession go parallel in many swamps. Some animals are manipulators of plant-succession processes. This chain has been disturbed. Turbidity in water has been detrimental to the growth of water plants, fresh and other aquatic animals. So the task to save the aquatic sites seems to be stupendous.

Obviously, the floral and faunal site identification and mapping of such areas involve a Herculean effort. The author enumerates the resources involved and sites endangered owing to biotic factors and analyse the difficulties in identifying such resources.

Wetlands in this country have shrunk considerably and there are endangered sites in all biotic communities and ecosystem. Highly productive wetland have been and are being destroyed for dredging, filling, mining, pollution, expansion of cultivation and rehabilitation.

What are Wetlands ?

Wetlands are economic reserves. They are for breeding, nesting, moulting, wintering by migratory birds. Such lands function as sponges accumulating excess water in flood and are important for the maintenance of ground water level. Marshes and swamps act as safety valves in the whole hydrographic system of a watershed area. Such lands are complete ecosystem and form vital sites for game birds of high genetic value. Ramsar conservation of IUCN had been an initial step to identify such endangered and rich sites of the world. This convention had emphasised that wetlands are vital links in the web of life and are sole habitats for several endangered species of local and migratory birds. Such sites have diverse, aesthetic and intangible values and provides fodder, forage and nesting sites for various avifauna and animals. Wetlands throb with faunal life and form nursery for very many aquatic animals and plants. Ponds are, however, the largest aquatic resources of the country.

Wetland definitions include three main components (Mitsch and Gosselink, 1986) which are :

Wetlands are distinguished by the presence of water

Wetlands also have unique soils that differ from adjacent upland

Wetlands support vegetation adapted to the wet conditions (hydrophytes) and conversely are characterised by an absence of flood–into lerant vegetation.

Ramsar Convention defines wetlands as ‘areas of marsh, fer, peatland or water, whether natural or artificial, permanent or temporary, with water, that is static or flowing, fresh, brackish or salt including areas of marine waters, the depth of which at low tide does not exceed six metres and may include riparian and coastal zones adjacent to wetlands, or islands or bodies of marine water deeper than six metres at low tide lying within. In addition, wetlands are ecotones since they are transition zones from uplands to deep water aquatic systems.

Values of wetland are, genetic conservation, water treatment, nutrient removal, fresh water fisheries, productivity, coastal fisheries, nurservices for ocean fisheries, tourism, coastal protection, energy and carbon dioxide storage, wildlife habitat.

Hollis et al. have produced a statement detailing the function of wetland and their human utilization.

WETLAND FUNCTIONS AND THEIR HUMAN UTILIZATION

Role	Elements	Function	Important to humankind	Unwise use
1	2	3	4	5
Store/sink	Rare, threatened or endangered plant and animal species and communities	Genetic diversity Recolonization source	Gene pool Science/ education Tourism Heritage	Excessive or uncontrolled harvest Damage removal or pollution
	Representative plant and/or animal communities	Ecological diversity Habitat maintenance	Gene pool Science/ education Tourism Recreation Heritage	Excessive or uncontrolled harvest Damage removal or pollution
	Peat	Nutrient, contaminant and energy store Habitat support Water storage	Fuel, Palaeo-environmental data Horticultural use Heritage Medicinal products	Drainage Harvest gaster than accumulation Destruction
	Human habitation sites	Archaeological remains	Heritage/ cultural Scientific	Destruction Lowering the water-table.

Role	Elements	Function	Important to humankind	Unwise use
1	2	3	4	5
			Recreation	
Pathway	Terrestrial nutrients, water and detritus	Food chain support habitat support	Food production water supply waste disposal	Interruption or abnormal change of flows pollution
	Tidal exchanges of water detritus and nutrients	Food chain support Habitat support Nursery for aquatic organism	Fish, shellfish and other food production waste disposal	Pollution Barriers to flow Dredge and fill
	Animal populations	Support for migratory species including fish	Harvest Recreation science	Overexploitation Interruption of migration routes Obstruction Habitat degradation
	Lakes and rivers	Waterways	Navigation	Obstruction Reduced flows and levels
Buffer	Water bodies, vegetation soils and depressions	Flood attenuation	Reduced damage to property and crops	Filling and reduction of storage capacity
	Water bodies, vegetation, soils and depressions	Detention and retention of nutrients	Food production Improved water quality	Removal of vegetation Drainage and flood protection
	Water bodies, vegetation, soils and depressions	Ground water recharge and discharge	Water supply Habitat maintenance effluent dilution River fisheries Navigation	Reduction of recharge Overpumping pollution

Role	Elements	Function	Important to humankind	Unwise use
1	2	3	4	5
	Water bodies and peat	Local and global climate stabilization	Equable climate for agriculture and people	Dessication
	Water bodies	Large volume Large area	Cooling water	Drainage Filling Thermal pollution
Producer	Production of plants	Food, materials and habitat for migratory species and grazing animals	Harvest of timber, thatch fuel and food science recreation	Overgrazing overexploitation drainage excess change to dry land or other agricultural uses
	Animal production	Fish, shellfish, grazing and fur-bearing animals	Harvest and farming	Overexploitation Excess change Habitat degradation
	Organic matter	Methane production Nutrient cycling	Fuel plant growth	Drainage Desiccation
Sink	Lakes, deltas flood plains	Sediment deposition and detention	Raised soil fertility clean downstream channels improved water quality downstream	Channelization excess reduction of sediment throughout
	Lakes, swamps and marshes	Bio-chemical self-purification nutrient accumulation	Natural filter for contaminants treatment of organic wastes pathogens effluents	Destruction of the ecosystem overloading of the system

But the author has not considered the aforesaid definition and have excluded brackish or salt or marine, riparian and coastal zones from the sphere of discussion in this chapter.

Threats to Wetland

Hollis et al. feel that wetlands cover 6 per cent of the world's land surface, and are found everywhere wetlands are under threat from agricultural intensification, pollution, major engenuring schemes and urban development.

Ramsar convention did not define 'wise use' nor identify ways in which could be achieved. The first conference at Cagliari (Italy) in 1980 approved a recommendation emphasizing that wise use of wetlands involves establishment to comprehensive wetland policies based on a nationwise inventory of wetlands and their resources. The next conference at Groningen (Netherlands) in 1984 adopted a recommendation which defined Action Points for Priority Attention in the field of wetland conservation. The Regina, 1987 (Canada) conference of the parties to the Ramsar Convention adōpted the following definition. The wise use of wetlands is their sustainable utilization for the benefit of human kind in a way compatible with the maintenance of the natural properties of ecosystem.

G.E. Hollis et al. in 'Wise Use of Wetland' published in Nature and Resources, January–March, 1988 have made the following comments on Wetland :

- Wetlands include some of the world's most productive ecosystem and they have a wide range of natural functions which are of value to humanity.
- Wetlands are also one of the most threatland habitats.
- Wetlands are regulators of hydrological cycle.
- Wetlands are nurseries and habitat for fresh water and marine fish.

Ralph W. Tiner (in Bio Science, April, 1991) in his paper 'The Concept of a Hydrophyte for Wetland Identification', has put much stress in the identification of wetland. He has redfined hydrophyte much different than those defined by Transleyer Weaver and Clemants; even trees have been included in the definition. Five definitions of hydrophyte have been drawn by Daubenmine 1968, Cowardin et al. 1979, Core of engineers 1987, Sipple 1988 and FICWD 1989.

Wetland in high altitude may be identified in Ladakh. The great salt lakes fringed with marshes such as Hanle, Harong, Chusul, Tsokar, Tsomoriri provide brief summer. In Kashmir valley wetland may be identified in Nowgong, Mirgund, Malgam, Haigam and Hokarsar. Keoladeo National Park at Bharatpur is an ideal wetland site that has abundance of bird population. The Rann of Kutch, a monsoon flood phenomenon which dries up into a flat desert at other times of the year, is crucially important as one of the major breeding habitats for the greater flamingo in Asia. Nal Sarovar in Gujarat is monsoonal that harbour migrant ducks and waders.

In the Ganga plan, mention may be made or marshy areas in North Bihar especially in Char areas. There are some sub Himalayan marshes as in Dudhawa National Park. Kabar Talin Bihar covers an area of 9000 ha. and support substantial population of migratory birds. It is said 70,000 birds are killed every year.

In the peninsular India there is dearth of wetland but during monsoon months there are plenty of wetland sites all over. Kollenru Lake in Andhra Pradesh is an ideal site for water birds.

Coastal wetlands are plantiful, but estuarine areas have not been treated as wetland in this study. Khijida area near Jamnagar, Ghed swamps at Porbunder, Point Calimere and Pulicat Lake in Tamil Nadu, Chilka Lake in Orissa and ideal aquatic floral and faunal sites.

Mapping and Selection of Sites

Thanks are due to the untiring and useful effort of the National Atlas Organisation to identify wetland of the country. They have also prepared forest cover and forest type map of the country on a broad outline. They have also marked the dams and reservoirs and wildlife sanctuaries including those on wetland.

6

GRASS AND MAN

An adage reads–"And I will send grass in the fields for thy cattle that thou mayest eat and be full."

Grass and grassland communities are an important natural resources. Grass as a family, Graminae includes a great variety of food crops, wheat, rice, maize, sugarcane, sorghum, millet, barley, oats etc. Sod crop is also included in it that provided forage or pasturage for grazing and domestic animals.

No appellations will be enough to express the multifaceted qualities of grass. "All flesh is grass," said a prophet. Grass has been considered as a gift of nature. battle fields strewn with dead bodies and saturated with blood, torn with ruts of cannon come to life again. Grass is immortal, while forests decay, harvest perish and flowers wither. Winter drives the shoot underground, which emerge at the first approach of spring. The tenaceous root system of grass conserve soil and water and reduce aridity in the environment; grass provide sustenance in one form or another for domestic livestock on which people depend for food and clothing. It is also needed as manure on cultivated field. Grass provided industrial raw material for paper and pulp. It is also used as straw board and for thatching rural houses. Grass is also the source of aromatic and edible oils, adhesives, plastic and starch. Grass invades the barrenness of deserts, climbs inaccessible mountains, amelioration climate and determine to a great extent history, character and density of nations. Grass is found in the wildest extremes of climate and it provides both staple cereal food (modified from J.J. Ingalls on Agriyenboot : 1948).

The primary form of food is grass. Grass feeds the animals; the animals nourish man; dies and become the grass again. Grassland characterises areas in which trees have failed to develop, because of the unfavourable soil condition, poor drainage and aeration, intense cold and wind, deficient moisture supply of repeated fire. Grass is admirably suited to withstand conditions of extreme moisture, excess drought and fires which would destroy growth of trees. Grass can be grotesque and vulgar; it can also be beautiful, ornamental and can be the source of endless quality material. Grass and grasslands control erosion, fix sand dunes, provide food and shelter to rodents, rabbits, avifauna and herbivorous. Run-off and loss of nutrients from agricultural fields, from farm and forest land are checked by grass cover.

If agriculture is to be the mainstay of our economy, then the importance of grass and grassland cannot be over emphasized. We need type of agriculture that is stable and secure for farms and farmers, an agriculture that can fulfil our needs of food, fibre, fodder and shelter for all times to come; and grass and grassland form an indispensable part of our agriculture.

Grass indicates the quality of soil and the types of forests. Specific species of grass determines the quality of timber species, their growth and changing status. Grass has many forms such as broad-leaved, thin-leaved, tall and short. Flowers of grass are produced in abundance which are smell and inconspicuous, massed in plumose or gracefully drooping panicles.

Bamboos are a variety of grass which have woody stem and petiolate or stalked leaves. Bamboos play a major role in the rural life and civilization of India. This country has 3.6 million hectares of land under bamboo with at least 136 species. Bamboos are poor man's timber and form an inexhaustible source of minor forest produce besides being the best source of raw material for pulp and paper.

India has many distinct patterns of grass distributed broadly over four zones of the country. These zones are the eastern, the western, the southern and the Himalayan. The eastern zone boasts of maximum number of species of grass. This zone has low and high level savannah like grass formation where many species of wildlife such as tigers, leopards, bisons, rhinoceroses, deer etc. occur. Khasi hills in Meghalaya offer extensive undulating grassland similar to those of Nilgiri downs. These grasslands are of biotic origin.

Unfortunately, in north east India vast areas of forests which have brought under shifting cultivation are now grasslands of various grades. The soil condition in these areas has undergone irreparable deterioration where nothing but grass grow. Many hill tops of Nagaland and Manipur are similarly grassy. In the western zone grass occurs over extensive areas, but the cover is very thin. Where over head trees exist the tree cover is also thin. The ground cover, however, offer shelter to very many species of wildlife. In the southern zone, Nilgiri downs are ecological wonders of Indian vegetation. The origin of the downs still remains a subject of controversy. Vast areas of dry deciduous forests have now been reduced to scrub savannah, shrub savannah and tree savannah as a result of forest depletion followed by grazing and fire. Many famous moist and dry deciduous forests where grass has held the ground have been conserved as wildlife sanctuaries and national parks. The Himalayan region has about 0.7 million hectares of alpine grassland between 3500 and 5000 metres. The origin of these grassland is non-biotic. These grassland assume colourful form owing to many herbaceous species coming into bloom in spring and in autumn. More than twenty wildlife sanctuaries and national parks have been established at this altitude. Rare species of birds and animals, such as snow leopard, blue sheep, musk deer, monal pheasant, snow cock etc. have found shelter in such sanctuaries. India is also fortunate in having about 1100 species of grass with more than a hundred distinct grassland types in the country.

Unfortunately, the research findings of the agronomists have not revealed the occurrence of highly nutritious and fast growing indigenous fodder grass in India. More than fifty species of exotic grass have been given trial in this country but results have not been satisfactory. One exotic grass–the Gustemele grass, has been widely cultivated in the foot hill areas of many Tea Estates of Assam and West Bengal for the purpose of soil binding, erosion control, nitrogen fixing, soil cover and as fodder. Mixed cultivation of grass and legume has not tilted expected results.

Various experts on grass, grassland and pasture have discussed in innumerable symposia, workshops and conferences on these subjects, volumes have been written and recommendations made; but only partial implementation of the recommendations could be made. The problems are mountain high. Uncontrolled grazing of 33 crores of cattle in this country has done irreparable damage to soil and has depleted environmental balance. For instance, the hilly areas of Himachal Pradesh, Uttar Pradesh and Jammu & Kashmir have 116,49,228 heads of cattle which have depleted forest cover over wide areas. In the districts of Chamoli, Almora and Chakrata in Uttar Pradesh there are 4.9, 9.0, and 7.7 cattle per hectare of land respectively which is more than the carrying capacity of the land. Besides, there are many nomadic graziers who have large flocks of cattle that graze in the forests. The forests of Himachal Pradesh have shrunk from 38 per cent to 14 per cent. Such irreparable damage from biotic factors of which grazing has played a major role has been supplemented by free grants of trees to the concessionists. These have also depleted the grassland. The similar situation occurs elsewhere in the country offering a bleak situation. Regeneration of forest crop has suffered. Softer species of grass have been eliminated and coarser species have survived as ground cover.

All these situations call for intensive attention and fact-based research on grass and grassland. It must not be under estimated that the prosperity and happiness of the country also depend broadly on cutious and scientific management of grass and grassland.

Ecological Status of the Grasses and Grasslands

There is no natural grassland of Savannah type in the states. Whatever grasslands exist today mainly result from fires either accompanied or not accompanied by shifting cultivation, grazing, and excessive cutting. The vast stretches of Savannah like formations that are found all over the states are certainly the results of fire, grazing and flood. If those limiting factors were to be withdrawn there would be a succession of vegetation towards an evergreen type. Innumerable indications of this can be seen in the forests. In the dry mixed forests, grasses occur in suitable pockets which are mainly caused by edaphic variations. Some of these pockets of grasslands have been there for about a century and there has been no large-scale invasion of evergreen or deciduous species succeeding them. The denseness of the stocking, thickly mated roots, annual fires and to some extent grazing have ensured this stability. In spite of these facts, the grassland of these regions is only relatively stable and cannot be called a climax formation. Such Savannah like formations are common in Terai and Bhabar areas, in river valley swamps and in burnt Sal forests in Assam. In the riverain forests of *Acacia catechu* and *Dalbergia sissoo,* tall grasses generally invade, and once burnt they persist, grow vigorous and suppress seedlings of the forest tree species. These conditions lead to the formation of vast areas of tall grasses in the riverain forests of terai and Bhabar zones. These grasses, however, are stages of primary succession.

In the alluvial laterite zones of West Bengal and Bihar there are no such extensive tract of grassland. In the laterite zones the soils are so poor that grasses sometimes fall to invade the eroded lands. There is a preponderance of annuals which come up in the rains and then disappear in the dry months. This is as a result of prolonged misuse of the soil by men succeeded by fire and grazing. The marshy areas of the alluvial zone bear coarse tall grasses. The fresh water swamps of Malda and West Dinajpur districts are notable as ideal habitats for many species of soft and coarse grasses Marginal lands are strongholds of tall grasses which invade and oust soft ones. This zone is thickly populated and forests and pastures hardly exist except in the extreme South. The human population and the vast number of cattle have suppressed the grasses which occur only sporadically. The downs of Meghalaya and of other hilly states have similar origin.

In the alpine areas of the Himalaya, extensive grasslands occur which are the result of devastating fire succeeded by grazing.

Thus it may be seen that the grasses and grasslands of the States are the results of biotic and edaphic factors leading to an edaphic or bioedaphic climax and there is no true climatic climax grassland.

Shifting Cultivation, Ecology of Grasses and Grassland

The shifting cultivation is an important agricultural practice of the hill regions in North-East India. It extends over 26000 sq.km. area in the states of Arunachal, Nagaland, Manipur, Mizorum, N. Cachar, Tripura and Meghalaya. Some areas have been cleared of forest trees for cultivation of Agricultural on a rotation.

The origin of the practice of shifting cultivation could be traced back to the Neolithic period about 7000 B.C. All hill tribes of India except the Apatains and Monpas of Arunachal Pradesh, the Angamis of Nagaland and the Hajomes of Garo hills (Meghalaya) are Shifting Cultivators.

Various stages of shifting cultivation are (*a*) selection of the spot for cutting trees and shrubs (*b*) process of cutting trees and shrubs in the spot (*c*) burning of plant material in site (*d*) cultivation on this spot for varying number of years depending on fertility of soil (*e*) abandonment of this spot for a limited period in favour of a fresh spot and after a lapse of 2–10 years reverting to original cleared areas.

The top soil is gradually washed away. Nutrients are leached and acidity increased. Micro-flora and fauna of the soil are disturbed. The tree vegetation are gradually exterminated. Long Jhuming interval allows time lag for tree species to grow to pole stage, but such intervals are gradually being cut down to minimum period owing to poor return of crop from the soil.

The Jhum cycles assumed by the North-Eastern council are : Arunachal, 2.69 year; Mikir hills (Assam), 7, 11 years; Manipur, 1.67 years; Meghalaya, 5.47 years; Mizoram, 9.74 years; Nagaland, 8.22 years and Tripura, 9.61 years.

Some villages in North Nagaland have been found to work on 2-3 years cycle. This situation prevails elsewhere also. Experts are of opinion that in Meghalaya, Mizoram and some of the states 64 per cent of the area in the hills are either barren or uncultivated. Obviously grasses have invaded such areas.

Repeated cutting, burning and ploughing with long Jhuming cycle have been responsible for profuse regeneration of bamboos (*Dendrocalamus hamiltonii, Melocanna baccifera, Neohouzeaua dullooa, Cephalostachyum pergracile, Bambusa pallida* grow over extensive area). Such bamboos are being exploited now to supply raw materials for various paper mills. Where Jhuming has been more frequent these secondary bamboos gradually vanished and gave way to tall grasses and reeds. Extensive areas of North Nagaland have no *Neyraudia reynaudiana, Thysanolaena maxima, Phragmites Karka, Saccharum* sp. which are being collected to feed Tuli paper mills. Pure or mixed formations of these species have been recorded over extensive areas.

Several stages of successions may be observed in Jhum areas. These are (*i*) pure and tall bamboo stage (*ii*) mixed bamboo and young broad-leaved pole stage (*iii*) tall grasses and young broad leaved poles stage (*iv*) short grasses and scattered young broad-leaved poles stage (*v*) short grasses and scrub stage (*vi*) eroded and barren hill slopes with sporadic grass and scrub (*vii*) eroded and barren hill slopes. Many species of softer grasses occur along with taller and coarser species.

Grass and Grassland (also bamboo) in various forests types of India :

(*i*) Bamboo brakes are often very dense. Throughout tropical evergreen forests in South India and Andaman the species viz. *Oxytenathera* and *Bambusa* species (Andamans) and *Ochlandra* a reed as South India are common in damp sites. *Ochlandra* is heavily exploited for paper and pulp industry. These also occur in semi-evergreen forests. A thorny bamboo (*Bambusa arundinacea*) is found along streams throughout India and after extends some distance up in the slopes in sheltered localities. North Bengal and Assam has a species *Dendrocalamus* hamiltonii which occurs on moist soil in moist deciduous forest over entire region.

(*ii*) Open Sal forests all over northern India have a heavy grass cover. These grass become tougher owing to annual fire which ultimately eliminates soft grasses and coarser grasses hold the ground (*Themeda* sp., *Imperata* sp., *Erianthus* sp., *Cymbopogon* sp., *Desmostachya* sp.,).

(*iii*) Although true savannah (derived from the Spanish sabana originally applied to the extensive, more or less flat grasslands or prairies of north and south America) does not occur in India. Tall grasses do occur all over India. Scattered trees of the early seres (successional stage of the normal succession on new reverain soil) stand over very dense tall grass growth of *Themeda, Imperata, Erianthus, Saccharum* etc. The riverain flats have stable cover of such grasses. These grasses become congenial niche for deer population where the predator tiger is also quite common. (Kaziranga, Manas, Jaldapara and many other wildlife Sanctuaries and national parks).

(*iv*) In mangrove, flats (newly formed) *Porteresia coarcltata* (Dhani grass) colonise as a pioneer vegetation which persists till such time when the broad leaved regeneration completely cover it.

(*v*) Most of the dry deciduous forests in India may be reduced to savannah form due to intense biotic interference. The grass becomes conspicuous whenever the canopy is broken and is favoured against other forms of vegetation by burning, lopping and grazing. Principal grasses are *Dichanthium annulatum. Bothriochloa pertusa, Eremopogon foveolatus,*

Heteropagon contortus, Cymbopogon coloratus, Themeda triandra, Arundinella setosa and *Sehima nervosum.*

(*vi*) *Pinus roxburghii* (Chinpins) forests are very frequently affected by fire and therefore, few tree or shrub species can grow in such forests. Such forests in the foot hills of Himachal Pradesh, Uttar Pradesh, West Bengal, Sikkim, Meghalaya, Nagaland, `Arunachal Pradesh have a thick undergrowth of *Cymbopogon, Heteropogon, Themeda, Arundinella* and several other grasses.

(*vii*) The temperate region of Nilgiri there are extensive undulating grassy hill hops which get burns periodically. Grading is also heavy in the grassland (Wenlock down). Species of grasses are *Bothriochloa pertusa, Ischaemum pilosum, Themeda cymbaria, T. Iriandra, Cymbopogon polyneros Eragr-ostis nigra* and *Tripogon bromoides.*

(*viii*) East and West Himalayan temperate coniferous forests have a wide number of species of *Arundinaria* which form dense, impenetrable brakes of one or more species over extensive areas.

Chakrata in Uttar Pradesh there are 4.9, 9.0 and 7.7 cattle per hectare of land respectively which is more than the carrying capacity of the land. Besides, there are many nomadic graziers who have large flocks of cattle that graze in the forests. The forests of Himachal Pradesh have shrunk from 38 per cent to 14 per cent. Such irreparable damage from biotic factors of which grazing has played a major role has been supplemented by free grants of trees to the concessionists. These have also depleted the grassland. The similar situation occurs elsewhere in the country offering a bleak situation. Regeneration of forest crop has suffered. Softer species of grass have been eliminated and coarser species have survived as ground cover.

All these situations call for intensive attention and fact-based research on grass and grassland. It must not be under estimated that the prosperity and happiness of the country also depend broadly on cautious and scientific management of grass and grassland.

7

REALISTIC CONSERVATION OF HIMALAYAN NATURAL RESOURCES

Conservation of the Himalayan natural resources and environment has been one of the major issues of the country to-day. Various institutions have been involved in the efforts, yet nothing much could be achieved. Of late, the Goverment of India has established a research and development institute under the control of Department of environment. The accepted fact of the day is that by conserving various natural resources it is possible to stop further depletion of the forests and land. But the dependence of the vast population on these resources is so great the complete ban on the felling of trees and exploitation of resources will only bring about economic crisis among the people. Conservation of flora and fauna, besides conservation of other natural resources will undoubtedly earn the region appreciable monetary benefits. In doing so, the people will be deprived of food, fodder, fuelwood, timber, resin and other minor products for their need. As such a workable balance has to be worked out, but how?

The Himalayas with their snows, glaciers, forests, rivers, avifauna, fauna etc. have been the source of inspiration of great poets and sages and they stand with their magestic and awesome stature to dazzle and daze the whole world. Due to the population explosion and consequent pressure on natural resources leading to over exploitation of forest resources in some areas, much of the resources has been depleted. The ecology, which has already been disturbed, stands the ominous risk of further degradation. The future of these natural resources look bleak and murky. Frequent occurrences of flood, drought and near-desert conditions stemming from the depletion of the resources are but nature's signals for greater disasters to overtake the country if we fail in our conservation effort. The great Himalayan ranges, 2400 km long and 240-400 km wide, with a number of peaks rising well over 7,200 m constitute the most striking feature in the geography of the world. The varied terrain range from lofty snow-clad peaks to deep valleys with detritus accumulation at the base of the foot-hills.

The Himalayan forests are diverse and comprise tropical, sub-tropical, montane temperate, temperate and dry and moist alpine forests and scrub land. There are many biotopes or life zones from tropical heat to alpine cold,–often telescoped within very short areas abounding in vast assemblages of flora and fauna unrivalled anywhere in the world. Dry and wet valleys add to the ecological complexities of the jumbled habitats rich in flowering plants, avifauna, insects, etc. The western Himalayan states have valuable coniferous (pines, firs and spruce) forests, the eastern Himalayas have, however, valuable broadleaved forests. All these forests are also very rich in fungi, lichen, moss and liverwort flora. In autumn and in spring miles and miles of hill slopes are found covered with red, mauve, blue white and cream coloured flowers that offer a wonderful kaleidoscopic view. The situation on the forest cover in western and central Himalayas is not at all adequate. About 50 per cent of the existing forests do not have adequate cover. But there are many reserves which are being carefully maintained. The forests in Bhutan, part of Sikkim and in Arunachal Pradesh are in pretty good shape. The Himalayan regions have about half the total number of the species of flowering plants, half of the number of the total avifauna besides rich assemblages of other fauna of this country. More than 550 ecological and climatic forest types in the region form special niches for myriad species of plant and wildlife. The floral diversity goes parallel with mammalian diversity. Quantitative estimates of the flora and fauna indicates this parallelism. Some revealing facts on these are : wide range of leaf surface, varieties of patterns of fruits and flowers, long duration flowers and fruiting, perallelism between the proportion of sizes of fruits and frugivorous avifauna, similar parallelism between insect size and the size of insectivorous avifauna, diverse floral and faunal elements of Africa, Europe, Chinese, Japanese, etc., very many mammalian species in high altitude and diverse genera and the species of various insect orders.

It is evident from the aforesaid brief synopsis that the Himalayan region is able to draw millions of tourists, besides botanists, zoologists, agronomists, geologists and other experts and conservation agencies that will enrich the economy of the region in various ways.

Now, let us take a look at the other side of the conservation problem. Records show that extraction of timber and fuelwood from the Himalayas is made to the extent of about 1.7 and 1.8 million m respectively. This fact alone will testify to inevitable dependence of people and industries on forest produce. Thousands of people are directly or indirectly dependent on various forests. Harvesting of timber and fuelwood will generate employment for 2.79 lakh man-days, involving a monetary benefit of Rs. 31.7 crores. Resin operation will generate an employment of 120 man-days per tonne. Minor forest produce will generate an employment of 0.18 lakh man-years. A complete ban on forestry operations will deprive the local people of all these benefits. These are, however, exclusive of the transport workers and material suppliers.

The very important aspect of newsprint and other types of paper needs to be discussed here. India import newsprint to the extent of 232 thousand tonnes per year (worth about Rs. 82.36 crores) which drains a lot of foreign exchange. India is already on the brink of paper famine despite the lowest consumption of 1.7 kg paper per year per capita as compared to 279 kg in the U.S.A. as such, it is imperative to work the forests on a sustained basis and bring vast denuded and blank areas under afforestation which again will generate employment.

Indirect benefit from working the forests cannot be ignored. The Central Government earns a lot in the form of taxes (Excise, Sales, Custom, Income, toll etc.) from industries (Paper, Board, Plywood, Matchwood, etc.) and other agencies. Besides these benefits, there are also many other benefits. Approximately about Rs. 700 crores circulate annually in the hills due to forestry only. Moreover, the socio-cultural, national defence benefits for keeping our northern frontier and border people gainfully and respectfully occupied is a benefit which cannot be measured.

While adumbrating a viable conservation policy for the Himalayan forest and other natural resources it should be kept in mind that the Himalayan resources cannot be exclusively used for aesthetic, cultural, educative, ameliorative, protective purposes leaving aside economic consequences. The fate of the vast number of people living in the hills is intimately linked with the exploitation of forest resources. Therefore, total ban on felling of trees or the exploitation of natural resources would be too disastrous for the Himalayan regional economy. The forest policy in the Himalayas has to be revised and modified in the light of the imperatives pointed out already. Forestry operations have to be done on a restricted scale. Resources have to be exploited without upsetting environment or economic balance. Afforestation have to be done on a realistic scale. Tourism should be carefully planned to protect the environment from depletion at high altitude. Millions of hectares (about 0.7 million) of alpine grassland which have a major role to play should be protected from the nomadic graziers. A sound conservation policy conducive to the needs of all interests concerned should be implemented.

Plant Resources

There are three major botanical regions. The western, the Central and the Eastern Himalayas. Exploration of plant species started as early as 1802 and is still continuing as there are still many unexplored regions. The conservationists have suggested creation of Biosphere reserves. The Government of India is actively considering to conserve flora, fauna and natural resources by creating wildlife sanctuaries, National parks, Biosphere Reserves and other protected areas a comprehensive list has been produced in the Chapter. India saves her wildlife. Entire region has plants and animals of phyto and zoogeographic significance owing to assemblage of various floral and faunal elements from Mediterranean and European to Sino Japanese and Malayan. The flora has high degree of endemism.

The famous botanist Janaki Ammal had found a high degree of polyploidy and identified the region as a region of active Speciation. This region has more than 8000 species of plants. Wide variability in the agri-horticultural wealth exist from the Sub-Himalayan to the Alpine region. The range and extent of genetic diversity of various crops differ within and among the western, the Central and the Eastern Himalayas. Some of the examples are Cereals (Wheat, Barley), Millets and pseudo cereals, Grain legumes and pulses, Oil seeds, Fibre crops, Vegetables, Spices, Condiments and baverages,

Fruits and Nuts, Medicinal and Aromatic plants. Medicinal plants cultivated in the hill and *Hyoscyamus, Datura, Physochliana, Chrysan themum, Dioscorea, Costos, Podophyllum, Digitalis Ephedra, Colchicum, Chirata* etc. Wild relatives of cultivated plants and allied types are *Solanum, Vitis, Saccharum, Dioscorea, Cucumis, Rubus, Pyrus, Prunus, Sobus* and *Cotoneaster.* In alpine pastures legumes and grass genera are *Caragana, Lespedaza, Medicago, Melilotus, Parochaetus, Trifolium, Trigonalia, Agrostis, Bromus, Dactylis, Agropyron, Poa* etc.

J.K. Maheshwari of N.B.R.I Lucknow has suggested 'Biotype Sanctuaries as an aid to the development of Himalayan Resources'. The Himalayas have been the theatre of evolutionary development of several angiosperms. There are regions of active speciation, hybridization and floral evolution. But the flora has been depleted considerably. Biotype sanctuaries representing gene pools of the major phyto-resources and wild resources and wild relatives or progenitors of cultivated plants, should be established in the region. He also mentions species noted in the previous paragraph. The concept of a biotype sanctuary recognizes the need for continued evolution of wild biota within natural environments so as to build up a wide genetic base.

P. Kapoor of Bio-Science Department, Shimla has listed wild plants used as food and suggests explorations of new food sources and make cyto-ecological, eco-physiological, nutritional and dietary evolution.

Animal Resources (mammals)

H. Khajuria studied mammalian fauna which shows tridimensional zonation i.e. longitudinal, latitudinal and altitudinal. Each major zone, a number of sub-zones can be recognized. The mammals show interesting adaptation and extra-limital affinities. It has been found that 40 species and sub-species have given rise to domestic breeds. Mammals of the Himalayas will now get protection in more than wildlife Sanctuaries, National parks and other protected zones.

The faunal elements in Western Himalayas comprise of peninsular, Indo-Chinese, Palaearctic and Tibetan. The peninsular element is found in forested foot hills *(Rhesus macaque),* Hanuman langur, the Bengal fox, the common palm civet, the small Indian civet, the grey mongoose, the ratel, the jungle cat, the leopard, the tiger, the wild boar, the four horned antilope, the blue bull, the barking deer, the hog deer, the swamp does Indo-Chinese element is represented by white throated martin, the serow etc., The plaearctic element is represented by wooly wolf, the brown bear, the pallas cat, the lynx, the snow leopard, the pine marten, the Siberian mink, the Urial, the pole cat, the Kiang, the Yak, the Tibetan element is represented hy Tibetan antelope, Tibetan gazelle, the markhor etc. The Government has established innumerable protection areas, laws have been enacted, but it depends on the people of the safety of resources will be ensured.

Live Stock Resources

In the Himalayas about 0.7 million hectares is covered by alpine pastures of good quality. Live stock population in 12 states comprise 20 million cattle, 10 million buffaloes, 3 million sheep, 6 million goat and 20 million poultry and ducks (Source : P.N. Bhat, I.U.R.I., Izatnagar). Dr. Bhat is of opinion that a Crop-live stock ecosystem mix which can maximise the yields of animal products, without disturbing the environment can be evolved.

Aquatic Resources

The aquatic resources has been discussed in the chapter."

Famous zoologist S.L. Hora remarked that the high velocity of river and stream water, water falls, rapid riffles etc. does not allow many species to survive. Despite the diversity of flora and fauna, there is ecological solidarity in having some species in Kashmir, Uttarkashi and in Arunachal viz., net-Spinning caddish (Hydropsyche), a swift water mayfly naiad (Heptagenia) and a water penny (Psephenus).

The golden mashcer, *Tor putitora,* and *Schizothoracid* species, *Schizothoraichthys esocinus* in different section of the Himalayan waters have become rare. Their preservation calls for complete ban on indiscriminate killing, forming of masheer, impounding of water in reservoirs, barrages, anicuts etc. prevention of organic pollution in rivers and lakes. Mortality of fish in Nainital lake was due to lower concentration of dissolved oxygen and the higher concentration of ammonia and hydrogen sulphide. This needs adoption of a drastic remedial measure.

Shifting Cultivation

Shifting cultivation, known as jhuming in the north-eastern region has affected the states of Arunachal Pradesh, Nagaland, Manipur, Mizoram, Meghalaya, Tripura and part of Assam. This has resulted in large scale deforestation, denudation of hill tops and slopes, soil erosion, setting up of reservoirs, streams and rivers leading to flood, loss of fertility of soil. This practice has directly affected all efforts of conservation of flora, fauna, water and land resources.

Woodbased Industries vis-a-vis Resource Conservation

R.K. Gupta of Central Soil and Water Conservation Research and Training Industries, Dehradun 1978 stressed the need of developing industry for sustenance of hill people in Garhwal Himalayas. The industries are the manufacturers of pencils, the matchwood industry, cheap furniture and agricultural implements, the sports-goods industry, the packing case industry, the wood-carving and engraving industry, the fibre industry, the paper-pulp industry, cultivation of drug and aromatic plants, besides live stock farming, but does not mention how these could be achieved without destroying natural resources.

Glacier Resources

C.P. Vohra of Geological Survey of India quotes Von Wissman who estimated that 33,200 sq.km. of the Himalayas and 16,000 sq. km. of the Karakoram is covered with ice. This area is 17 per cent of the Himalayas and 37 per cent of the Karakoram respectively. The Sutley basin had 6390 sq.Km. under ice cover (11 per cent of total basin area). Vohra estimated figure for the total number of glaciers in the Himalayas is around 15,000.

The glaciers are an important source of fresh water stored in the solid state. Von Wissman estimated the volume of ice in the Himalayas to be 1400 cu km. The glaciers, glaciated regions and the snow cover exercise a crucial influence on many vital elements of the Himalayan environment in it's climate, in water resources, its debris and situation phenomena, floods, vegetation cover, communications and safety of the structures.

V.S. Krishnaswamy, Ex-Director General of G.S.I. stressed the importance of earth resources (water resources, mineral resources, geo-thermal resources, hydro electric resources and land resources) of the Himalayas.

Water Resources Conservation

The Himalayas are the source of three major river systems – the Indus, the Ganga and the Brahmaputra. The Indus has a catchment of 468,000 sq.km. upto Indo-Pakistan border; of this 321,000 sq.km. lies in India. The Ganga with its tributaries has a catchment of 511,000 sq.km., out of which 300,000 sq.km. lies in India, while the Brahmaputra has a catchment of 580,000 sq.km. of which 187,000 sq.km. lies in India (Y.K. Murti, C.W. Commission 1978).

The Irrigation Commission in 1972 determined the irrigation potential of the three rivers as Indus (6120 ha. thousand), Ganga (27350 ha. thousand) and Brahmaputra (1710 ha. thousand).

Murti (CWC) considers measures to develop water resources to be taken to meet various demands in next few decades. Unless proper corrective measures are taken, the construction of works to develop the water resources potential could lead to far-reaching changes in the Himalayan eco-system such as in sedimentation, soil erosion, stratification etc.

8

CONSERVATION OF MANGROVE FORESTS FOR HUMAN WELFARE

Coastal woodlands, known also as mangrove forests (or as littoral, estuarine or tidal forests) are nature's masterpieces of creation and ecological bonanza. These are, according to a UNESCO expert "Unique resources that are of great importance to the majority of people of this planet."

The inhaustible potential resources of the forests did not receive the attention either of the foresters or the administrators. During the last decade only the scientists assembled in various symposia and workshops and discussed about various resources. potential and problems of these forests. In 1978 a working group on mangrove (term used by all international organisations) was created at UNESCO's initiative. This working group in its first meeting resolved to undertake a biosphere inventory of mangrove lands and to determine their current status. In 1980 the Commission of Ecology (COE) of International Union for Conservation of Nature and Natural Resources (IUCN) created a working group on mangrove ecosystems to collate existing information. The Department of Science and Technology, Government of India formed a task force to evaluate the mangrove forest potentials of the country; later, the Department of Environment, forest and wildlife has been involved in conservation aspects of Mangrove.

Special features of mangrove forests that set them apart from other types of forests are:

* the trees and bushes grow below the high tide level,
* the root system is inundated by saline water at regular intervals,

* the forests are favoured by tropical humid climate accompanied by silt-laden rivers that form mud flats, although some grow on extremely arid coasts and on coral and rock islands,
* the trees develop special adaptation mechanisms such as breathing roots of various types, fleshy and thick leaves, salt glands, viviparous germination, sunken stomate, aqueous tissue etc.

Population pressure combined with major and multiple demands on the coastal resources for rehabilitation, urbanisation, creation of ports, recreation facilities, aquaculture has caused substantial damage to such forests locations of mangrove forests.

Unlike various types of tropical and temperate forests, mangrove forests are located in small areas covering about 84000 sq. ha. along the coasts in most of the tropical maritime countries. The largest single patch of mangrove forests is in the Sunderbans of Bangladesh and India, which are idiosyncratic in their formations, physiographic characteristics, peculiar floral and faunal assemblages and in the behaviour pattern of resident animals. A huge population of over three million people depend directly or indirectly on these forests.

India has only 6,81,976 hectares of mangroves forests of which 4,18,828 hectares occur in the Sunderbans and another 1,15,000 hectares is in Andaman and Nicobar islands. The remaining 1,47,888 hectares occur in the deltas of Mahanadi, Gadavari, Krishna and in Saurastra and Kutch. Except in the Sunderbans, the forests elsewhere are in a deplorable state. This calls for immediate rehabilitation of these forests. The mangroves in Andamans have also suffered from heavy felling for collection of fuelwood.

The Dynamic Nature of Mangrove

Estuaries are complex biological sites which are very rich. Here the opposite currents meet and oscillatory tidal currents are formed which make the sedimentary system complicated. Physical, chemical, mechanical and animate environments play a crucial role in modifying activities of microbial organism and other animate objects which are linked by processes resulting in a continuous exchange and assimilation of energy. The 'internal' process of energy fixation, accumulation of biomass, decomposition of dead material and mineral cycling are strongly influenced by a small set of critical 'external' processes which affect the availability of water and essential nutrients. The principal characteristics of the coastal environment are both cyclical and unidirectional because of intense interaction between land and sea. The meeting zones of fresh and sea water generate a ecosystem where nutrients are trapped. Food web structure in coastal zone is extremely complex. Benthic micro-algae and organic detritus (seat of algal micro-organism) constitute the principal energy source for vast population.

Mangrove Resources

Mangrove forests are characterised by a very low floristic diversity. There are about 90 plant species of which 55 are restricted to mangrove swamps. The indo-pacific region is very rich in number of species and the quality of these forests is very high. Excellent quality charcoal is produçed from *Rhizophora* (Gurjan) which has a heavy demand. Firewood is produced from many species and is heavily exploited. *Heritiera* (Sundri) of Bangladesh is the most valuable timber of the mangroves. *Excoecaria* (Gnewa) is exploited in Bangladesh for pulpwood and matchwood; elsewhere it is used as packing case timber and as fuelwood. *Xylocarpus* (Pussor) is used for house post. Tannin obtained from *Cerops* (Goran) and *Rhizophora* (Gurjan) are used in leather, ink, plastic and formal-dehyde industries. Various parts of mangrove plants are also eaten. The fruit of *Avicennia* (Bean) and *Sonneretia* (Keora) are edible as are the seedlings of *Avicennia.* Radicles of *Bruguiera* (Kankra) are eaten. Salads are made from the leaves of *Sonneretia* (Keora) and from the herb *Sesuvium* (Gimasag) considerably quantity of honey and wax is collected from Sunderbans. *Nipa* (Golpata) is a very useful plant of the mangroves. Lower surface of young leaves is used as cigarette wrapper, while the older leaves are used for thatching purposes. The leaves also serve as raw material for paper industry. The sap of the inflorescence can be tapped for sugar, alcohol, vinegar production while its tapping gives a juice and a whisky is produced from it.

The mangroves support and sustain marine fisheries by serving as nurseries and as feeding ground for many marine invertebrates and fish by producing large amount of detritus for the marine food web. These forests serve as a link between terrestrial and marine ecosystem which is highly productive. Large primary production is consumed directly by animals, the major part entering the food web as dead organic matter and detritus being consumed within the mangrove. The dead leaves are fed upon by crabs and other animals. As the leaves have high cellulose content they are only partly digested by the animals and are broken into small pieces. These broken pieces provide substratum for bacteria and fungi. This process of partial digestion and microbial growth is repeated which constitutes the basis for the detritus food web. Innumerable species of crustaceans, molluscs and fish take part in the degradation of dead plant tissues.

Part of the life cycles of many marine organisms is spent on the mangroves. Many species of *Penaeid* shrimps spawn off-shore using mangrove as a refuge and feeding ground in later stages of their life cycle. It is not known if the "mangrove dependent" species only require a bottom mud with certain salinity regime or are influenced by other factors. Shrimp and fish population migrate from place to place and it has been found that there is a positive correlation between shrimp landings and mangrove forests; also such correlation exists between landings of penaeid shrimps and areas of intertidal vegetation. The large fresh water prawn *Macrobrachium* species which reach a length of 30 cm. passes through their initial stages of the life cycle in brackish water and later wander upstream to fresh water areas. Mud crab-*Scylla serrata*, the delicious and highly priced crab, form a highly delectable item of food. Cultivation of this and other crabs in the Philippines, Thailand and in Singapore is done extensively by private agencies. mangrove oysters are highly marketable edibles. Various snails, clams, and cockles are collected from mud flats. Even the mud-skippers fetch high price in Hongkong and in the Philippines. In the Sunderbans there are more than 87 species of fish a hood number of which form commercial species. The gulf of Mexico provides classic of estuary dependent fisheries. About 90 per cent of the commercial catch and 70 per cent of the recreational catch in the area are dependent on the estuarine habitat for part or the entire life cycle. More than 40,00,000 hecters of tidal areas in southeast Asia has been brought under brackish water pond culture.

Myrids of species and individuals of invertebrates, their specialised life-cycle, colouration, and behaviour pattern in coastal and sandy shore are specific qualities of the mangroves. Of the mammals the occurrence of man-eating Royal Bengal tiger and their mysterious behaviour pattern are wonders and myths of animal world. Pigs and spotted deer also occur in abundance. Estuarine

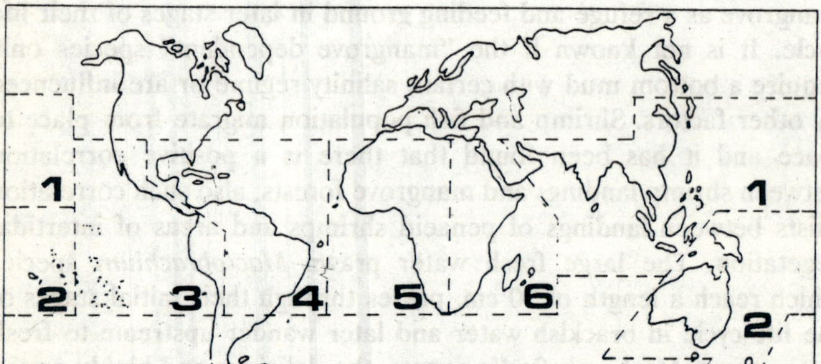

Fig. 8.1
Source : IUCN Com. on Ecol. No. 3, 1983

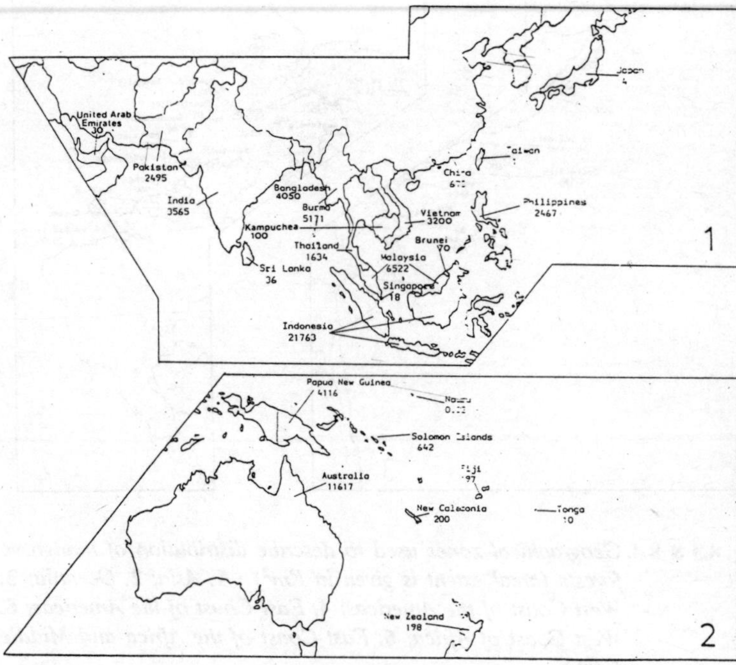

Fig. 8.2
Source : IUCN Com. on Ecol. No. 3, 1983

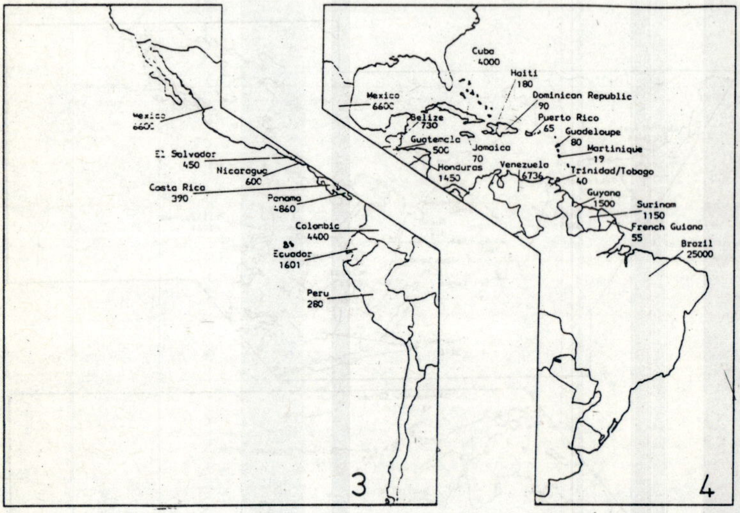

Fig. 8.3 & 8.4. *Geographical zones used to describe distribution of mangrove forests (areal extent is given in Km*^{2}*) :* **1.** *Asia;* **2.** *Oceania;* **3.** *West Coast of the Americas;* **4.** *East Coast of the Americas;* **5.** *West Coast of Africa;* **6.** *East Coast of the Africa and Middle East.*

Source : IUCN Com. on Ecol. No. 3, 1983

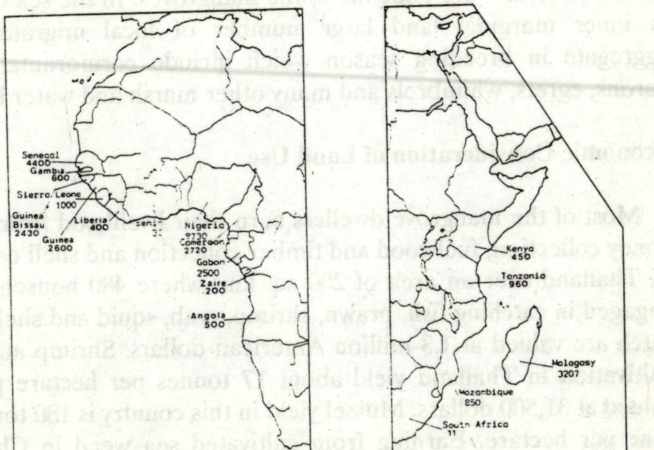

Fig. 8.5 & 8.6. ***Geographical zones used to describe distribution of mangrove forests (areal extent is given in km^2) : 1. Asia; 2. Oceania; 3. West Coast of the Americas; 4. East Coast of the Americas; 5. West Coast of Africa; 6. East Coast of the Africa and Middle East.***

Source : IUCN Com. on Ecol. No. 3, 1983

crocodiles (*Crocodylus porosus*) are the largest and most ferocious of all crocodiles. Monitor lizards, rhesus monkeys, fishing cats, dolphins and sharks in water are few, amongst many, animals of the mangroves. proboscis monkeys are endemic to Borneo. Crab eating macaques, *Presbytes* langur and flying foxes are other animals of the mangroves. In Papua New Guinea spotted Cuscus are very common. Of the birds, few are endemic to the mangroves. In the sea coast and in inner marginal land large number of local migratory birds aggregate in breeding season which include cormorants, darters, herons, egrets, whimbrels and many other marsh and water birds.

Economic Consideration of Land Use

Most of the mangrove dwellers earn their livelihood from fishing, honey collection, fuelwood and timber collection and shell collection. In Thailand over an area of 200 sq. km. where 480 households are engaged in catching fish, prawn, shrimp, crab, squid and shell annual catch are valued at 1.3 million American dollars. Shrimp and oyster cultivation in Thailand yield about 17 tonnes per hectare per year valued at 31,500 dollars. Mussel yield in this country is 180 tonnes per year per hectare. Earning from cultivated sea weed in China and Korea is high. In the Philippines a sea weed farm of 0.5 hectare has provided a net income of 1360 American dollars. In some of the southeast Asian countries milk fish and shrimp catch from intensively managed area is over 200 kg. to 3000 kg. per year per hectare.

Mangrove forestry, capture fisheries and open water aquaculture are largely complementary activities which are one form of land use. Removal of mangrove forests adversely affect capture and open water aquaculture. This indirectly means disappearance of mangrove dependent species. The highly productive and dynamic sites can be very usefully utilized by judicious management to uplift the standard of living of the poor mangrove dwellers. The following will testify to high potential of mangrove sites.

Charcoal, Nipa thatch, small scale fishing, oyster cultivation and shrimp cultivation may yield value of 318, 230, 39, 60 and 206 American dollars per year per hectare respectively; in a commercial farm shrimp cultivation has yielded a sum of 2106 American dollars per hectare per year where rice cultivation in the mangroves yielded an income of 165 American dollars.

In terms of gross annual income per hectare the best land use appears to be intensive shrimp cultivation. Charcoal also yields a very high value (a plantation may yield a value of 400 American dollars per hectare per year).

All these indicate that there must be judicious management of the productive mangrove site as all the processes and animate objects are intimately connected with each other.

Conservation and Human Welfare

Besides utilizing the resources for monetary gain there is urgent need to protect the forests from conversion to agricultural land and fish-prawn breeding sites. Sincere effort, however, has been made to preserve the mangrove ecosystem as a whole in Australia where 23 'Habitat Reserves' have been marked covering an area of 93,925 hectares; but mention must be made of Kakadu National Park in northern territory which gives protection of about 614,400 hectares. Bangladesh and India afford protection of about 150,000 hectares of mangrove forests in the Sunderbans. Brazil, Colombia, Costa Rica, Dominican Republic, Malaysia, Mozambique, New Guinea and many other countries have felt the necessity of preserving such forests. These sites have provided immense recreational and aesthetic facilities to the people, besides affording research facilities on unknown plants and animals and on their behaviour pattern. The utility of the mangrove forests to counteract the force of cyclonic wind and waves has been felt in India, Bangladesh, in the Philippines and else where. The destruction of these forests may eliminate very many species of fish, shrimp and prawn besides other species which spend either entire or part of their life cycle in the estuaries. Acid sulphate that is produced in soil where tree felling has been done is indicative of misuse of the mangrove productive sites.

Yet extensive destruction has been done to such forests. 1300 to 1500 sq. km. of forests have been reclaimed in Indian Sunderbans in the last 100 years. More than 40,00,000 hectares of tidal area in Asia has been brought under brackish water and pond culture. About 27000 hectares of mangrove land in peninsular Malaysia has recently been identified for brackishwater fish culture. Biotic interference has depleted the mangrove forests all over India except the Sunderbans.

The foresters, technologists and economists must be involved in preparing master plans for best utilisation of the mangrove forests. It is indeed a very difficult task as to how much area is to be set apart for forestry, aquaculture, conservation reserve and for other purposes. Careful considerations must be made to utilize the mining sites. The population in the vicinity of mangrove forests is very high and they are susceptible to various contagious diseases. Enormous number of vectors are associates with aquatic environments. As such the planners have a very arduous task to perform in order to utilise the vast resources to the benefit of the people. Few areas in tropical forests can provide such varied types of resources as the mangrove forests and proper management of these forests is sure to bring health and wealth to the mangrove dwellers.

The causes and consequences of Mangrove destruction :

Two maps are produced (Source 'Global Status of Mangrove Ecosystem' 1983) show the countries and areas covered by mangrove vegetation.

The causes of mangrove destruction as currently underway in the World can broadly be distinguished as :

(1) over-exploitation by traditional users; and
(2) destructive action resulting from activities generally unrelated to sustained uses of mangroves.

These causes can be further subdivided by the scale of impact likely to result from a single decision. The scale of impacts commonly encountered with destructive uses is listed in the Table.

It is apparent in this classification that repeated or simultaneous actions in a region increase the total impact. For example, one traditional exploiter is insignificant, however 10,000 exploiters focusing on one area would have a significant impact. Also, a combination of actions imposed in a local area would have a cumulative impact on the total system.

Scale of Impacts

Activity	Scale of Impact ha
Clearfelling	10,000 - 500,000
Diversion of freshwater	1,000 - 500,000
Conversion of agriculture	100 - 100,000
Conversion to aquaculture	100 - 10,000

Activity	Scale of Impact ha
Conversion to salt ponds	100 - 1,000
Conversion to urban development	100 - 1,000
Construction of harbours and channels	100 - 1,000
Mining/mineral extraction	10 - 100
Liquid waste disposal	1 - 10
Solid waste/garbage disposal	1 - 10
Spillage of oil and other hazardous chemicals	1 - 10
Exploitive traditional uses	1

Mining/Mineral Extraction

Mining of alluvial mineral deposits in the tropical coastal zone is on the increase. Tin, Chromium, Titanium and some other minerals are exploited. Such mining results in complete destruction of forests. The dominant effect in the deposition of overburden materials which are transported to and within the mangroves by surface water. Excessive sedimentation, turbidity of water caused by dredging and overburden disposal results in destruction of local corals and sea-grass meadows and their associated faunas. Mining activities may require various refining processes to further concentrate the one for economical shipment. Such procedures may involve crushing, washing, chemical floatation, screening and dewatering. Dilling for production of oil occurs in Mangroves.

Diversion of Freshwater

Mangroves are not salt demanding plants. They best develop in areas which have significant input of fresh water run-off especially in arid and semi-arid areas. In considering the major rivers which have been dammed (e.g. Nile, Indus, Ganga etc.) the down stream effects on both coastal vegetation and nearshore fisheries, will become readily apparent. The diversion of freshwater from the mangrove ecosystem affects mangroves and associated fauna in a variety of ways. The dominant effects results from progressive increases in dry-season soil salinity and result in the gradual displacement of mangrove species by others more tolerant of the increased salinity. This can have severe consequences when a local industry is dependent on a sustained supply of the species being replaced, for example, the commercial mangrove forest tree, Heritiera fomes in the Gangetic delta of Bangladesh.

Reduced dry-season fresh water in the mangrove environment also affects terrestrial fauna dependent on both a source of freshwater and sufficient food during the dry season. The aggravated effects on the Bengal tiger (*Panthera tigris tigris*), for example, have been documented, and similar information is available for other large animal residents of Asian mangroves. In the past in many tropical countries, natural mangrove forests have provided sanctuary for many species, simply due to their existence in a coastal landscape otherwise modified by economic development.

High water salinities affect mangrove-dependent fisheries. The reduced flushing may result in the accumulation of detritus where it is unavailable to dependent off shore communities. Too many barrages constructed on the Indus has resulted in less discharge of freshwater for about nine months as a result of which mangroves have vanished.

Bangladesh

The mangrove in the delta of Ganga–Brahmaputra has been in a state of progressive deterioration. Salinity gradient controls the structure and composition of forests as may be seen in the Bangladesh (less salinity) and West Bengal (high salinity). Inspite of the fact that the Sundarbans Forest is a major source of round wood and pulpwood in Bangladesh and is a major wildlife refuge for certain rare and endangered species (e.g., the Bengal tiger *Panthera tigris tigris*), only minor attention has been given to the deterioration on the forest. Further, it is doubtful that continuing economic development based on fresh water resource development will be sacrificed or altered to protect the Sundarbans. However, co-operation between India and Bangladesh has resulted in a greater dry-season flow of freshwater to south-western Bangladesh.

Forest Exploitation

Mangrove forests are much used for fire wood (India & Africa),

Charcoal (Thailand, Malaysia and Indonesia). Timber is also exploited all over for construction, pulp and paper. Large scale settlement of people by felling mangroves has been done in Venezuela and East Kalimantan (Indonesia). Large scale deforestation has also been made in Sarawak, Sabah, Matang, Venezuala, Sutawesi, North Sumatra and other areas.

Conversion to Agriculture and Aquaculture

Coastal swamp forests in saline, anaerobic environments have traditionally been considered to be marginal or totally unsuitable for agricultural and/or aquacultural production. The Rice Research Institute in Los Banos, Philippines find ways to utilize this marginal resource which includes not only mangrove dominated lands but also contiguous areas of tidal flats, beach ridges, inter creek, saline flats and reclaimed lands. In Asia and Africa the pressure in arable land has led to conversion of mangrove forests into agricultural land. This has been done in Senegal, Gambia, Sierra Leone, India and Bangladesh.

Aquaculture projects have expanded enormously in recent years resulting in conversion of 1.2 million ha of mangrove forests (1977). In Philippines aquaculture area was 174,101 ha. in 1972. As of 1981, there was demand for additional 70,300 ha. of mangrove areas. Ecuador and Costa Rica mangroves have also suffered such consequences.

Coastal Development

Coastal development for tourism, urban area extension etc. has taken place in Australia, New Zealand, U.S.A. and in other countries. Singapore has also suffered a lot of reduction of area.

Salt Pond Construction

Salt ponds are built on mudflats or more commonly in mangrove areas. In Gujarat there is a huge Chemical-Salt Industrial complex. In Benin, Malaysia and some other countries there are salt pond areas.

Construction of Channels and Harbours

Harbour construction has taken place in China, Australia, Japan,

Panama, Costa Rica and some other countries which has affected mangrove forests to a considerable extent.

Solid Waste Disposal

Owing to heavy growth of population the garbage and solid waste have increased several-fold. The leachate of materials, plus any toxic materials included in container cause pollution. In Brazil solid wastes of 130 Tons per day is being deposited in Hacorobi mangrove swamp. Similar incidence is observed elsewhere.

Liquid Waste Disposal

Waste materials which are toxic to the living organisms is generated in most of the industrial production processes, whether agricultural, forestry, chemical or mining industry. Urban and domestic human activities also generate effluents containing substances with potential toxicity. Extensive work is being undertaken to classify phyto and zoo-plankton species in the estuarine areas according to their tolerance to toxicity in estuarine water.

Oil and Other Hazardous Chemicals

Oil spill is common in port areas due to transport of bulk liquids, particularly crude and refined petroleum products. Owing to the pressure of mangrove–dominated shore-lines close to major shipping corridors, mangroves are more frequently being impacted by accidental or wilful spillage.

TABLE: *Plants and animals at risk*

Species	Common name	Status
PLANTS AT RISK		
Bruguicra gymnorhiza	Mangrove	Extinct–Taiwan
Bubophyllum avicella	Orchid	Believed extinct–Singapore
Bulbophyllum blumei	Orchid	Endangered–known only from one or two localities in Singapore and southern Johore

Species	Common name	Status
Bulbophyllum hotryphora	Orchid	Endangered–known only from one or two localities in Singapore and southern Johore
Bulbophyllum concinnum	Orchid	Endangered–known only from one or two localities in Singapore and southern Johore
Ceriops tagal	Mangrove	Vulnerable in Taiwan where only one stand remains Vulnerable–South Africa
Dendrobium cruentum	Ueang nok kaeo	Endangered–Thailand
Dendrobium flavidulum	Orchid	Endangered–known only from one or two localities in Singapore and southern Johore
Dendrobium flexible	Orchid	Endangered–known only from one or two localities in Singapore and southern Johore
Dendrobium kelsalii	Orchid	Endangered mangrove form in Singapore: the mountain form is fairly widespread
Dendrobium pensile	Orchid	Endangered–known only from one or two localities in Singapore and southern Johore
Eria pudica	Orchid	Endangered–known only from one or two localities in Singapore and southern Johore
Finlaysonia obovata	Mangrove liane	Vulnerable–occurs only in Malaysia and Sundarbans but only grows in virgin mangroves, not in regrowth area
Grammatophyllum speciosum	Queen Orchid	Endangered–Thailand

Species	Common name	Status
Lumnitcera racemosa	Mangrove	Vulnerable–South Africa
Oberonia flabellifera	Orchid	Endangered–known only from one or two localities in Singapore and southern Johore
Paphiopedilum exual	Lady Slipper	Endangered–Thailand
Plocoglottis lowii	Orchid	Vulnerable–occurs only on sandy soil in Borneo, Sumatra and Malaysia at landward mangrove margin
Rhizophora mucronata	Mangrove	Endangered–Taiwan
Schoenorchis perpusillus	Orchid	Believed extinct–Singapore
ANIMALS AT RISK		
Mollusca		
Strombus gigas	Botuto	Vulnerable Venezuela
Terebralia palustris	Mangrove whelk	Vulnerable South Africa
Reptiles		
Alligator mississippiensis	American alligator	Threatened Florida, USA
Caiman crocodylus	Baba	Endangered–Venezuela
Crocodylus acutus	Gran caiman/ American crocodile	Believed extinct Venezuela Endangered–Florida, USA
Crocodylus porosus	Saltwater crocodile	Vulnerable; protected within reserves–India and Malaysia; collecting prohibited–Australia, India and Sri Lanka
Drymarchon coarais couperi	Eastern indigo snake	Endangered subspecies Florida, USA
Epicrates striatus fosteri	Bimini boa	Vulnerable–found only in Bahma Islands

Species	Common name	Status
Nerodia fasciata taeniata	Atlantic salt-marsh snake	Endangered subspecies Florida, USA
Birds		
Agelaius xanthomus	Yellow-shouldered blackbird	Endangered–USA
Agelaius xanthomus xanthomus	Puerto Rico yellow-shouldered blackbird	Vulnerable– found only in Puerto Rico
Amazona arausiaca	Red-necked parrot/Jacquot	Endangered found only in Dominica
Amazona vittata	Puerto Rican parrot	Endangered found only in Puerto Rico; protected within a reserve
Anas acuta	Pin tail duck	Endangered–Puerto Rico
Amazona bernieri	Madagascar teal	Vulnerable–found only in Madagascar
Anas clypeatra	Shoveler	Endangered–Puerto Rico
Anas crecca	Green wing teal	Endangered–Puerto Rico
Anas platyrhynchos	Mallard	Endangered–Pucrto Rico
Ardea cinerea	Grey heron	Endangered–Malaysia
Ardea herodias	Great blue heron	Endangered–Puerto Rico
Ardea purpurea	Purple heron	Endangered–Malaysia
Ardea sumatrana	Dusky-grey heron/ Great-billed heron	Endangered–Malaysia
Aythya collaris	Ring neck duck	Endangered–Puerto Rico
Bebrornis sechellensis	Seychelles brush warbler	Vulnerable–found only in Seychelles; protected within a reserve; collecting prohibited
Botaurus lentiginosus	American bittern	Endangered–Puerto Rico
Casmerodius albus	Garza real/Royal heron	Vulnerable–Venezuela
Catoptrophorus semi-palmatus	Willet	Endangered–Puerto Rico

Species	Common name	Status
Circus buffoni	Long-winged harrier	Vulnerable–Trinidad and Tobago
Coccyzus melacoryphus	Dark-headed cuckoo	Vulnerable–Trinidad and Tobago
Cochlearis cochlearis	Boat-billed heron	Vulnerable–Trinidad and Tobago
Columba leucocephala	White crown pigeon	Endangered–Puerto Rico
Dendrocygna arborea	West Indian tree duck	Endangered–Puerto Rico
Dendrocygna bicolor	Fuluus tree duck	Endangered–Puerto Rico
Dendroica petechia petechia	Barbados yellow warbler	Endangered subspecies Florida, USA
Dichromanassa rufescens	Reddish egret	Endangered Puerto Rico
Dryolimnas cuvieri - aldabranus	Aldabra white-throated rail	Vulnerable found only in Aldabra; protected within a reserve
Dupetor flavicollis	Black bittern	Endangered Malaysia
Egretta alba	Large egret / Great egret	Endangered–Malaysia and Puerto Rico
Eudocimus ruber	Corcora / Scarlet ibis	Vulnerable Venezuela, Trinidad and Tabago
Falco peregrinus	Peregrine falcon	Endangered–Florida, USA
Fulica caribaea	Caribbean coot	Endangered–Puerto Rico
Halcyon senegaloides	Mangrove king-fisher	Vulnerable–South Africa
Haliaeetus leucocephalus	Bald eagle	Endangered–Florida, USA
Ibis cinereus	Milky stork	Endangered–Malaysia
Ixobrychus cinnamomeus	Common bittern	Endangered–Malaysia
Ixobrychus involucris	Stripe-backed bittern	Vulnerable–Trinidad and Tabago
Laterallus jamaicensis jamaicensis	Black rail	Endangered–Puerto Rico
Leptoptilus javanicus	Lesser adjutant stork	Endangered Malaysia

Species	Common name	Status
Limnodromus griseus	Short-billed dowitcher	Endangered Puerto Rico
Limosa fedoa	Marbled godwit	Endangered Pucrto Rico
Limosa haemastica	Hudsonian godwit	Endangered–Puerto Rico
Mycteria cinerea	Milky stork	Vulnerable–protected within a reserve on Pulau Dua off western java; collecting prohibited in Malaysia except by aborigines
Numenius phaecopus hodsonicus	Whimbrel	Endangered–Puerto Rico
Nycticorax nycticorax	Black-crown night heron	Endangered–Malaysia and Puerto Rico
Ortalis vetula deschavenseei	Utila chachalaca	Endangered–found only on Utila Island in Caribbean
Oxyura dominica	Masked duck	Vulnerable–Trinidad and Tobago
Oxyura jamaicensis	Ruddy duck	Endangered–Puerto Rico
Pandion haliaetus	Aquila pesadora/ Osprey	Endangered–Puerto Rico Vulnerable–Venezuela
Paroaria gularis nigrogenis	Red-capped cardinal	Vulnerable–Trinidad and Tobago
Pelecanus occidentalis	Brown pelican	Endangered–USA
Phalacrocorax carbo	Common cormorant	Endangered–Malaysia
Plegadis falcinellus	Glossy ibis	Endangered–Puerto Rico
Pluvaialis dominica	Golden plover	Endangered–Puerto Rico
Podiceps dominicus	Least grebe	Endangered–Puerto Rico
Porphyrula martinica	Purple gallinule	Endangered–Puerto Rico
Porzana carolina	Sora rail	Endangered–Puerto Rico
Porzana flaviventer	Yellow-breasted crake	Endangered–Puerto Rico
Pseudocolopteryx sclateri	Crested doradito	Vulnerable Trinidad and Tobago

Species	Common name	Status
Sterna albifrons	Least tern	Endangered–Puerto Rico
Sterna hirundo	Common tern	Endangered–Puerto Rico
Sterna maxima	Royal tern	Endangered–Puerto Rico
Sterna sanvicensis	Sandwich tern	Endangered–Puerto Rico
Xiphorhynchus picus altirostris	Trinidad straight-billed wood-creeper	Vulnerable–found only in Trinidad; protected within a reserve
Mammals		
Canis rufus	Red wolf	Believed extinct–Florida. USA
Cyclopes didactylua	Silky ant-eater	Vulnerable–Trinidad and Tobago
Felis concolor coryi	Florida panther	Endangered subspecies Florida. USA
Felis pardalis	Ocelot	Vulnerable–Texas to Argentina; collecting prohibited in most countries of its range
Macaca fascicularis	Long-tailed macaque	Protected within reserves Malaysia
Nasalis larvatus	Proboscis monkey	Found only in Borneo; protected within reserves
Odocoileus virginianus clavium	Key deer	Endangered subspecies–Florida, USA
Panthera onca	Jaguar	Vulnerable–South American countries; protected within reserves in Colombia, Peru, Bolivia and Brazil; collecting prohibited within most of range
Panthera tigris sumatrae	Sumatran tiger	Vulnerable–found only in Sumatra; collecting prohibited
Panthera tigris tigris	Bengal tiger	Endangered--protected within reserves in India, Bangladesh, Bhutan, Burma and Nepal

Species	Common name	Status
Preshytis cristata	Leaf monkey	Some geographic subspecies are vulnerable–Thailand
Procyon cancrivorus	Crab-eating racoon	Vulnerable–Trinidad and Tobago
Pteropus vampyrus	Malaysian flying fox	Protected within reserves Malaysia
Sciurus niger avicennia	Mangrove fox squirrel	Found only in southern Florida where two breeding populations remain; protected under Florida legislation
Trichechus manatus	Manatee	Endangered Venezuela and Florida, USA

Source : Global Status of Mangrove Eco-saplem, 1983 (UNESCO)

9

ENVIRONMENTAL HAZARDS ON TRIBALS IN FORESTS

Tribal environment is a complex and a multifacet subject all parameters of which are not closely interwoven. It concerns 'Environmental impact' which is the identification of basic data about external environment such as climatic data, surface water flow pattern, local flora and fauna etc. "Environmental Impact' study is involved in assessment of hazardous and to a limited extent "Environmental analysis". Secondly, social forestry and tribals in India need be interwoven as most of afforestation and rehabilitation in Forestry Sector is being done under social forestry. According to accepted definition (F.A.O. 1979 and a National Commission on Agriculture 1976) Social Forestry includes any situation which initially involves local rural poor in forestry activities which embraces a spectrum of situations ranging from woodlots to growing of trees in farm land; thirdly, many tribal communities are forest dwellers-whose social, cultural and physical well-being centres round the forests. The forests have of late, suffered a large scale degradation and the tribal people have been exposed to various of socio-cultural and environmental hazards.

So this study has to list the diverse nature of environmental hazards that the tribal themselves and the habitat have suffered and how far forestry and Social forestry practices can restore and ameliorate various environmental, socio-economic, socio-cultural changes.

The choice of the subject has been unique and appropriate. The Forest (Conservation). Act, 1980 and National Forest Policy 1988 have specifically acknowledged the rights and concessions of the tribal people in clear terms to protect their interests, elevate their standard of living and improve the quality of life. This effort is well supported by Social Forestry which has become a major element in India's overall programme of rural development. As Environmental Impact Analysis (EIA) for drawing up Environmental Management Plans (EMP) are considered imperative before implementation of any major project now-a-days, the study is likely to be of considerable significance by providing data base facts for use in such quantitative assessment in future.

India is in the Realm of Environmental Crisis

The agencies like "The Brandt Commission", "The Global 2000 Report to the President of USA" and "The World Conservation Strategy" have indicated the environmental crisis in developing countries. The Planning Commission document "Report of Committee on Forests and Tribal of India 1982 enumerates India's dressed environmental crisis that the country is reeling under flood, droughts, soil erosion, forest and land depletion, that there is famine in the supply of fuelwood and timber. The report of National remote Sensing Agency (NRSA) revealed about alarmingly low forest cover (only 14 percent) and that 1.5 million ha. of forests are being depleted annually. Shifting cultivation in Tribal areas is contributing substantially to this crisis.

Tribal Demographic Scenario

Anthropometric information on the tribals of India in the tropical forests is inadequate. A satisfactory assessment of the environment and the socio-economic and working condition of the population is yet to be made. There is need for extensive studies in physiological characteristics of physical work capacity and heat tolerance, about their migration, adjustment, physical adaptation, method of food intakes, nutritional status etc. for proper assessment of tribal environment.

Many people living in or in the vicinity of forests, in recently deforested areas, many have modified or radically altered their living condition changed with either improvement or worsening hygiene and be exposed to different infections. Such data would be necessary for proper formulation of development programme and environmental assessment. The investigation requires careful planning, clear aims and interdisciplinary approach.

The tribal dwellers have suffered heavy stress due to deforestation and depletion of forests, raising of monoculture crop, reduction of food sources (plant and wildlife). There has been an increasing geographic, social and demographic scale of economic system including the development of large extractive industries and destructive environmental modification for the benefit of societies geographically or culturally based largely outside the tropical forests. So there has been an abrupt change in population-environment relationship specially in use-system which are sensitive to population density, or hunting and gathering. Increased migration has caused cultural heterogeneity. Owing to opening of mines and industries in the vicinity of tribal areas they have been suddenly exposed to a changed physical, chemical and biotic environmental stresses; sudden alteration of diet might have caused profound disturbance of water and electrolyte balance and caused various diseases. The migration to a new environment might have deleterious effect on Social and economic status including poor housing, sanitation, water supply and exposed to many infections diseases.

A close look at the forest, shifting cultivation, tribal and mineral maps of India will reveal that the tribals from an integral part last two areas. There are more than 400 different ethnic communities forming about 38 million population (1971 Census) which increased to 42 million when the area restriction relative to the demographic distribution of the Scheduled tribe population within a state was removed. It forms about 6.64 percent of the population of the country. Nearly 82 percent of this population is concentrated in central and western part of the country, about 11 percent in the north-eastern states and about 7 per cent dispersed in small pockets in the southern zone. State wise distribution in Madhya Pradesh (20.13 p.c.), Orissa (23.11 p.c.), Assam (10.99 p.c.), Andhra Pradesh (3.8 p.c.), Bihar (8.35 p.c.), Gujarat (13.99 p.c), Maharashtra (5.86 p.c.), Rajasthan (12.13 p.c.), Meghalaya (80.43 p.c.), Nagaland (88.76 p.c.), Tripura (29.98 p.c.), Arunachal Pradesh (78.85 p.c.), Mizoram (94.28 p.c.), Manipur (31.13 p.c.) and West Bengal (5.72 p.c.) The Planning Commission document (1981) considered the hill area of U.P., West Bengal, Assam as backward areas though exclusively backward areas are hill states of Jammu and Kashmir, Himachal

Pradesh, Manipur, Nagaland, Mizoram, Maghalaya, Tripura and Sikkim also the hill areas above 600 m in the Deccan belt has been considered as backward areas.

Some principal tribes are : Bhils of Gujarat and Rajasthan, Nagas of Nagaland, Gond/Agaries of Madhya Pradesh, Khasis of Meghalaya, Santhals of Bihar and West Bengal, Khariars of Bihar and Orissa, Savaras of Orissa, Lepchas of Sikkim, Todas of Tamil Nadu, Konda Beddis of Andhra Pradesh, Kolams of Adilabad, Nissis of Subansiri and several others. Environmental Impact Assessment (EIR)–In Tribal areas a Complex Study :

What is Environmental Impact ?

There are few fields of activities as pervasive and as controversial as environmental impact analysis. Environmental impact can be defined as any change in the environment that is caused by an activity or a factor which may be physical, chemical, biological social or economic.

An environmental impact analysis is a study of probable changes in the various Socio-economic and Biophysical characteristics of the environment which may result from a proposed or impending action. (R.K. Jain et al. in 'Environmental Impact Analysis').

To accomplish this analysis, it is first necessary to develop a complete understanding of the proposed action. What is to be done? What kind of material, Manpower and/or resources are involved. Second, it is necessary to gain a complete understanding of the affected environment. What is the nature of bio-physical and/or socio-economic characteristics that may be changed by the action. Third, it is necessary to project the proposed action into the environmental characteristics, quantifying the changes whenever possible. Fourth it is the analysis of the probable environmental consequences of the proposed action may be used in the decision making process. The exact procedures to be followed in the accomplishment of each E.I. Analysis are by no means simple or straight forward.

Environmental Impact Assessment

Assessment as defined in Webster's New Twentieth Century Dictionary of the English language is 'an evaluation', further, to assess is to "........fix value of........" In essence, an assessment is a quantitative evaluation for the purpose of establishing relative values among various quantities of parameters. In environmental assessment, parameter are chosen to reflect the quality of environment and the effects favourable or unfavourable, of proposed actions on the quality. Thus, environmental assessment consists in establishing quantitative values for selected parameters, which indicate the quality of the environment before, during and after the action.

E.I. Assessment is the documentation of an environmental analysis which includes identification, interpretation, prediction and mitigation of impacts caused by a proposed action and project. This document (EIA) is utilised to provide a basis for intra-agency review of project impacts and is designed to provide adequate information for judging whether an environmental impact statement should be prepared. It is the most powerful tool hitherto available to the environmental managers for anticipating and averting environmental damage. Several Techniques of varying complexity have been devised and used as guide to decision makers.

Evaluation Methodology

The assessment of impact brought about by a large number of actions in the tribal habit and habitat which are physical, chemical, biological, social and economic etc., is extremely complex. It is more difficult to establish quantitative value for selected parameters in such assessment. However, common methods put forward by Leopold, I.B. et al. in the article "A procedure for evaluating environmental impact" (published in Inf. Cir. U.S. Genl. Survey, 1971) are generally tried. There are various other methodologies which need careful study before they are applied.

Complexity of Study in Tribal Areas

These are due to–close correlation between the tribals cultural attainments and habits with the habitat they live in, their life-style and existence depend on favourable physical environment (which also include their house food tools etc.), their evaluation has parallelism with floral and faunal evolution in the same habitat for centuries, These are from four major stocks of mankind : Negroid,

Astraloid, Mangoloid and Caucasoid and have three major families of languages viz., the Austro-Asiatic, the Dravidian and the Tibeto-Burman, there are various patterns of societies, villages, marriage types, ethnological groups, nomadic groups, they have occupational diversity and diverse source of economy, they have symbiotic relationship with forests, they are primitive societies which have achieved some kind of adjustment between its material needs and the potentialities of its environment, and they have several sizes of social group, varying material needs and various degrees of skill to tap the resources.

Moreover, plants are interwoven with rites and rituals. Many of the leaves, fruits etc. needed in magico-religious rites, are procured from forests. Many indigenous forests are considered abode of dieties and are worshipped. All these make any assessment extremely difficult.

Environmental Impact of Tribal Renaissance on Socio-Cultural Scenario

Several tribal areas have undergone administrative changes. Reorganisation of states took place in several states. Also ecology, economy, social systems, religion, value systems suffered some change. Development of road, concrete house, railways airports, motor cars, cinema houses, schools, colleges, health centres, factories, mineral and non-mineral mines, welfare centres etc. have affected the root of socio-cultural scenario. Vast number of mines were thrown open and expanded. This brought employment, improve communication system and welfare facilities and change in the pattern of living. Owing to growth of population, there has been increased requirement of fuel wood, bamboo and timber and this led to ecological depletion and dilution of diversity and density.

With the awareness about ecology, the clamour for stoppage of Jhum cultivation increased and schemes were started to regulate and replace Jhum cultivation for which land capability and soil survey studies have been undertaken. Barren areas were brought under coffee, rubber, tea, pineapple and various other horticultural crops. All these activities are fast changing the life style of the tribal people. This change has been faster in north-eastern states owing to more spread education and more awareness of a better future. Various Schemes have been introduced by the Government to improve this socio-economic conditions of the tribals and therefore society structures are also fast changing. The sixth schedule allowed the formation of District Council run by educated persons. This spread more education more civic facilities were extended. During the second and Third Plans the development of villages and small scale industries became the land mark of industrial development.

The rapid changes as detailed in last two paragraphs might have brought about undesirable cultural disorganisation and some sort of resistance against industrialisation which need a thorough assessment.

Impact Due to Transmigration

India is intent on developing the country's isolated and alien groups. To achieve that end the tribals sometimes are resettled under transmigration programme. This is to raise their living standard and incorporate them; into the main stream of Indian Society. But this may mean their alienation, poverty and cultural death. They are being cut out of the social and cultural fabric that has given their lives meaning for thousands of years.

In funding various transmigration schemes the World Bank has breached its own economic and environmental policies in some countries. Some environmentalists linked this issue in Indonesia as an environmental impact to "Waging the equivalent of thermo nuclear War" and "as the Single Sectoral activity with the greatest potential to advance forest destruction–often to no constructive results". The Dandakaranya Project was human and natural disaster where the settlers never settled properly and the tropical forests of Baster, Korpat and Dhenkanal districts lost species diversity, exposed sterile, soil caused species extinction and nutrient deficiency of soil. A vast area of 75,000 sq. km of Tribal reserve where the Gonds, Khonds, Jalapas, Bhuiyas, Battadas tribes used to live was destroyed. These tribes must have faced a large scale impact on their family, education, caste, attitude health, religion, consumption pattern, food and migration pattern.

In erstwhile Bhopal State rehabilitation of Bhils, Gonds, Korki, Keer, Mongra, Sansia tribes were made. Many Landless were settled. Similar rehabilitation was done in the states of Baster, Sarguja and Jashpur. The displacement of tribals (28, 275 families) from 216 project areas during Second and Third Five Year Plan involving 4210 ha. of land. B. Chowdhury in his book "Tribal Development in India". The Bhillai Steel Plant, the Korba Thermal station, the Parasia Coal mine and the Kolma Collieries are located in tribal areas and the tribals had to migrate elsewhere. Several Chinchu tribals are to be shifted from Amrawati Tiger Reserve.

Impact on Ethnobiology

Ethnobotany is defined on "Systematic study of the botanical knowledge of a social group, its use of locally available plants as food, medicine, clothing and religious rituals". It is also the study of cultural use of plants. Ethnozoology has the similar definition. The studies made by Botanical Survey and Zoological Survey of India have recorded wide scale depletion of such resources. Several Research Projects have seen initiated by the "Ministry of Environment and Forest", such studies have revealed that a veritable arsenal of wild plants with potential for exploitation of humanity is available in the forest of which the tribals aboriginals have specific knowledge. Over 1900 plants and animals had been recorded in 1984 used by the tribals for food, fodder, fibre, medicine, clothing, shelter etc. The Z.S.I. has recorded mammals and birds used as medicine by the tribals; they use blood, bone marrow, liver, urine, excreta as medicine.

The conservation of plants and animals of the forests had been a part of their faith and traditions and a set of fables, folklores, totoms, toboos, religious belief perpetrated this culture.

Impact of River Valley Projects

In 1987 the Government of India granted the states of Madhya Pradesh and Gujarat permission to transform the Narmada River and it's valley through a series of 3000 associated dams, 135 minor and 30 major dams. The principal focus of attention was the two giant reservoirs, the Sardar Sorobar in Gujarat and the Narmada Sagar in Madhya Pradesh. About 3,50,000 ha. of forest land and 2,00,000 ha. of cultivated land would face irreversible environmental changes. The tribals residing in the area are the Bhils, and Gonds and the Banigas. About 10 lakh people would likely to be affected of which substantial population would be tribals who have to be transmigrated. The Bodhaghat Hydel Project in Baster would affect about 9000 tribals who have to be transmigrated from the submerged area (13,526 ha.).

The Munda tribals are fighting against Koel-Karo project in Chotanagpur plateau. Munnar in Kerala, Bedhhi in Karnataka, Tehri and Vishnu Prayag in U.P., Lalpur in Gujarat are other projects but fewer tribals will be disturbed. Propatpatnam and Inchampalli in M.P. Maharashtra-Andhra Pradesh border will drown about 2,00,000 ha. of land and affect 40,000 media Gond tribes who would have to be transmigrated to Marathi speaking area, Baba Amte considers this as Cultural Ethnocide".

About Dihang a project in Arunachal Pradesh the Chief Minister remarked "We cannot agree to a project like Dihang as it would affect the life style, culture and environment of almost one sixth of the population of this sensitive area".

Impact of Shifting Cultivation

Ower more than 1,00,000 sq. km. of forested land in the states of Orissa, Madhya Pradesh, Andhra Pradesh, Bihar and all the states of north eastern India have suffered from vegetation and land degradation, and depletion of water regime. With the pressure of population jhum rotation has become shorter and shorter and has led to soil deterioration, loss of species diversity, depletion of wild life and eventually loss of ecological balance. According to Planning Commission (1982) 233 blocks spread over 62 districts in 16 states the jhum is practiced involving 12 p.c. of tribal population.

Names of the tribes in various states involved in jhum area :

Andhra Pradesh	–	Bagata, Gandabas, Khonda, Koyas, Konda Beddis and others
Assam	–	Mikir, Mizo, Naga, Chakma, Jainlia,
Bihar	–	Malpaharia, Birjea, Korwa
Gujarat	–	Bhil, Kunbi, Kokanis
Kerala	–	Paniyan, Irular, Kurumbas
Madhya Pradesh	–	Agariya, Baiga, Bharia, Gond, Kowra, Media etc.
Maharashtra	–	Major gond, Thakur, Kunbi, etc.
Orissa	–	Bhuiya, Gadabas, Kond, Koya, Kotia, Saor, Saora, Paroja.

Districts other than North-eastern India where jhum cultivation is done are–Andhra Pradesh (East Godavari, Srikakulam Vishakhapatnam) Bihar (Senthal Parganas), Madhya Pradesh (Baster), Orissa (Dhenkanal, Ganjam, Kalahendi, Keonjhar, Koraput, Sambalpur and Sundergarh).

Impact of Mineral Environment on the Tribals Living in Forests

Although mines of India occupy an insignificantly small areas the impact of mines and mineral based industries in tribal belt is significant and it is an imperative obligation of the administrators to restore, rehabilitate and revegetate such areas by biological and engineering processes.

A superimposition of three maps, viz. forest, tribal inhabitated and mineral will testify to the fact that mineral and tribal areas are interwoven. As such, this insignificantly small area (relating to the geographical area of India) inhabitated by various tribal communities from a carcinogenous nucleus in a tribal environment. Some observations are as follows :

(*a*) *Subarnarekha Basin*

The basin, the smallest of fourteen river basins of India, is a typical example of hazards brought about by mining on a good section of 11 lakhs scheduled tribe and 9 lakhs scheduled caste.

Once a quite coy with cystal-clear stream blooming with lotuses hid behind primeval forests, the subarnarekha is now rocking under booms of dynamite charges and the rattle of heavy earth moving machines, continuously unearthing its rich mineral wealth and demolishing the remaining forest cover soil mantle all over the basin is eroded.

(*i*) Strip mining, mine tailings, ore drumps of copper mines, radio active waste materials in the uranium mines at Jaduguda contaminate ground and surface water in tribal areas.

(*ii*) Processing of Uranium ore which are converted into magnesium diuranate by wet process contaminates more areas.

(*iii*) The Apatite (which contain fluorine) ore causes fluorine contamination in surface and ground water system.

(*iv*) At Rukha (near Ranchi) Subarnarekha contains 0.74 milligram of fluoride per litre of water.

(*v*) Gorumahishni–Badampahar ranges have a large number of abundance mines, unprotected high benches on the hill slopes are polluting both land and water at lower level.

(*vi*) Quarrying of China clay, limestone etc. also cause wide scale pollution.

(*vii*) Analysis made under the auspices of Global Environmental Monitoring System (GEMBS) at Ghatsila and Talisilwai in Subarnarekha show low (2 mg per litre) Dissolved oxygen content.

(*viii*) Copper deposits have other associate elements such as Sulpher Nichel, Cobalt, Molybdenum Selenium, Gold, Titnium, Silver, Platinum and Uranium. Rain water percolating, through the open sulphide ore dump and huge dumps of rock wastes leach toxic elements lime, slecniu, molybdenum and Uranium. Some toxic elements find way to agricultural land.

(*ix*) Vanedium at Dublabera, Kumbhar dubi, Bisai (Mayurbhanj District) Chromite at Roruburu, Chilungburu, Kimsiburu and jojohatu China clay between Kharkai and Vaitarani Basin put substantial land out of use. Tailing of asbestos mine does not bear any vegetation.

(*x*) Chromium as Chromite has minimal toxicity, but should it be oxidised to hexavalent Chromium the tailing could become very toxic and may be carcinogenic.

(*xi*) Iron and Manganese oxide play an important role in the soil in fixing trace elements such as Cobalt, Copper, Zinc, Nickel as well as pollutant like lead. The associates of these elements with iron and Manganese has important implications for agriculture and plant growth. Once fixed, the elements are unavailable to plants. High pH, excess of phosphate bicarbonate, copper, zinc, cobalt, cadmium, manganese or nickel in the growth medium may cause iron deficiency.

So the tribals residing in the Subarnarekha basin have to use contaminated water, their agricultural land has been rendered unfit for use and workers face noise and air pollution.

Tribals in Orissa Mines

Orissa has 237 mines (Keonjhar 66, Sundargarh 81, Cuttack 16, Sambalpur 38, Mayurbhanj 18, Other 18) and about 800,000 people are employed there. An investigation in 28 mines in Keonjhar and Sundergarh districts showed about the tribal engaged is about 85 per cent. Safety rules are not observed meticulously and the mines are exposed to various types of hazards such as noise, dust and polluted water. Besides, a good types of hazards such as noise, dust and polluted water. Besides, a good percentage of Mayurbhanj's 58 percent, Sundergarh's 53 per cent, Keonjhar's 46 per cent, Koraput's 56 per cent of Tribals are exposed to hazards of mining.

Tribals and Pollution

Ferro-Manganese Plants at Rayagoda and Joda, Pig iron plants at Joda, Ferro-Silicon and Silicon metal at Theruvali, Cement plants at Bargarh, Fertilizer plants at Rourkela, Thermal power plant at Telcher, Sponge iron plant near Palaspanja are some industries in Orissa that are causing hazards in tribal areas. Many such instances can be stated. Salenium is present in sulphide material, coal, phosphorites and uranium bearing rocks and soil derived therefrom. Selenium in human system is responsible for certain acute and chronic diseases in man and animal. Selenium rich zones are Singhbhum, Aravalli group of rocks, Meghalaya, upper Assam, Damodar Valley, Jabalpur, Mahanaeti Valley, Satpura hills and Guddepa basin. The toxic effects of Selenium are concern in tribal areas.

The tribals in Central India, whose lands and forests have been extensively destroyed by mining and associated industry are worst sufferer. No less than eight super Thermal Power Stations have been planned in a cluster to the coal mines of Singrauli. Similarly all major steel plants are concentrated in Bhilai, Bokaro-Rourkella-Jamshedpur Durgapur Burnpur region. Chhattisgarh was once a rice bowl with fertile land now poverty striken people move about under oppressive contractual condition. About 2,50,000 tribals are engaged in Mica mines who brave various hazards of mining. 20,000 workers at Cement factory at Khinkpani, Chaibasa where Hoand Diker tribes reside mostly. Chittaranjan locomotive work has Santhals and kol tribes residing in the vicinity of the workshop. Narmada valley has rich coal, iron, limestone, copper, dolomite, manganese, bauxite and soapstone material which are tribal areas. A lot of tribal workers are engaged as unskilled workers in various industries at Amar Kantak, Bharat Aluminium company, Hindusthan Aluminium Company etc. In Balaghat district Malanjkhand Copper mine point is causing serious ecological problem. 10,000 sq. km. of coal areas and 14,000 sq.km. of bauxite area cause various hazards to tribals. Rajasthan accounts for lead, zinc, tungsten, gypsum, asbestos, etc. have been responsible for soil salinity due to mining.

Impact of Air Pollution on Tribals

Several instances have so far been cited about the hazards of various mining operations faced by the tribals. Wind carries dust from waste heaps, blasting and heavy machinery-operated carries dues area. Blasting produced noxious fumes. These produce respiratory diseases and eye ailments. Phenumoconiosis, Silicosis and Asbestosis are caused by inhalation of dust. Common disease that follow are Cataract, Conjunctivitis, Corneal ulcer, Glaucoma and Squint trachoma.

About Roro asbestos mine at Chaibasa I.L.O. in 1978 remarked "the Rore mine endangered not only the lives of mine workers but those living in the surroundings areas". Ore concentrators, crushers and smelters are further sources of pollution.

Impact of Land

Agricultural land degradation has been mentioned in para 6. Limestone quarries and cament factories have reduced and productivity in areas of Udaisagar, Khamli and Chittor in Rajasthan, Agricultural fields in a 10 mile radius around the cement factory at Khikpani in Singhbhum lie barren under layers of accumulated dust.

Impact of Water Pollution

Ample instances have been cited in para 'D' on water pollution.

Ore fines and toxic substances carried by raid water into water courses alters their chemistry and make it unfit for human consumption. The untreated effluents, slime or taiting are released neighbouring stream.

- Effluents from Bailadilla Iron ore mines containing 18 per cent of mined ore are dumped into nullah. The turbid streams join the Sankini river turning its water red. This water is used by a population of 41 villages with 40,000 villagers mostly tribals who have named the river as "The river of Sorrow".
- Continuous dewatering under ground mines (millions of litres) cause flooding, silting, water-logging and pollution. Water table is lowered.
- Underground waters of Birsinghpur (M.P. Sadole pump out 3000 litres of water per minute into Ganyra nullah. The villages of Nayveli lignite mines have protested against such dewatering.
- The Chotanagpur plateau and the Maikal Ranga feed five Major rivers–the Narmada, Sone, Rihand, Mahanadi and Damodar and the area is vulnerable.
- At Baster the National Mineral Development Corporation (NMDC) may get lease 800 ha. of Kanger Resource for Dolomite. This may cause concern to local tribals.
- The relation between agriculture and trace metals is to be solved as elements like Cadmium or Lead which does not have biolgical function is potentially harmful.

Social Forestry Ameliorate Hazards in Tribal Areas

The aim of Social Forestry is

- To meet the most urgent requirement of fuel wood in shortest possible time and to provide poles, timber, bamboo, fodder, fruits and other minor forest products for basic requirement of rural population.
- To provide employment to the unemployed and under employed local people, perticularly the landless agricultural

labours including tribals, scheduled cast and other traditionally weeker sections of the rural communities.

-- To induce community participation in creating, maintaining and protecting plantation programme under a joint management system with share of benefits of the produce raised.
-- To improve the environmental condition by creating green belts in urban and industrial areas thereby ensuring aesthetic and recreational facilities.
-- To supply improved variety of seeds, grafts, seedlings for increased production and liftment of the economic standard of local people.

Various objectives have met with commendable success in West Bengal. Similar success has been achieved in Gujarat, Uttar Pradesh, Tamil Nadu and some other states. But the aspect of improvement of environmental condition in creating green belt in urban and industrial areas leaves much to be desired. Some tangible benefits have occurred due to planting on non-forest land where soil water conservation status are improving the quality of life.

Lately, there has been a new stratagy to allow the participant people part or full share of the yield. Generally the following components are built in any social forestry programme -

(i) Reforestation of degraded Government Forest Land.
(ii) Community forestry in barren I and H, other Government Deptt. Land, Panchayat land, Institutional land etc.
(iii) Farm Forestry on privately owned waste land.

This has created an atmosphere when more and more people are being motivated and involved in the Social Forestry Programme.

Social Forestry in Tribal Areas

Social Forestry has so far not been a promising tool for socio-economic and ecological development, continuous employment, for better nutrition, higher quality of clothing, home, better health and recreational facilities in tribal areas. Bulk tribal population has not been covered under direct and immediate micro-level benefits to resource–poor and economically disadvantaged rural dwellers, as well as bring forth long term, macro-level, resources-conservation type advantages that Government agencies try to achieve. Neither, it has proved useful, as a small-scale landuse operation ranging from pure forests to integrated agro-forestry in tribal areas; nor it has initiated inter-action between the forest ecosystem and human social systems in tribal inhabited area. It has yet to deal with broad aspect of human ecology in tribal areas and will need to consider social factors such as cultural knowledge and values regarding forestry, availability of resources land, capital, materials and labour, social constraints on resources management, social competition and conflict over resources. It is considered a major element in India's over-all record development but it has to cover bulk rural areas of central India.

Need for Planting Tribal Inhabited Mine Areas

The problems of dust, mine wastes, acidity, toxicity, presence of trace elements, heavy metals carcinogenic materials and other pollution hazards in mines areas of about 6000 mines have been identified. The role of vegetation to mitigate such hazards has been considered an imperative tool considering its comparatively low investment and lasting benefit. Let us enumerate in start the benefits of tree planting which are as follows :

- -- Dense vegetation intercepts incoming and outgoing rediation and dusty wind, screen air particles, create a mucroclimate and ameliorate weather besides supplying fuel wood, fodder, timber, fruit and M.F.P. Such screening is absolutely necessary in mining areas.
- -- One ha. of spruce collects about 32 tonnes of dust from atmosphere, a Beach tree collects 68 tonnes. So creation of barries of dense tree crop supplemented by shrubs and herbs is essential. Such planting will purify air in mining area.
- -- Dense ground vegetation (grass, agave, herbs shrubs) arrests flow of water and help deposition of silt and reduce erosion and water pollution hazards. Leaves will absorb sunlight, reduce ultraviolet rediation.

-- The vegetation will screen field and pastures against redioactive dust to save people from the effects of radio activity contaminated nutrition.

-- The sound absorbing quality of the trees will depend on its structure (depth, thickness and types of leaves etc.). The denser the undergrowth and the ground vegetation and conopied nature of forests, the greater its vertical closure, and higher in its sound absorbing capacity. A dense vegetation of 50 m width or varying height reduces the traffic noise by 20-30 decibels.

-- A. Bernatsky's work on the role of protective forests in industrial site, may be considered for reducing hazards surroundings of industrial areas where the tribals live.

-- Plants with a high metal or chemical tolerance can be effectively used as indicator and pollution scavenger which have developed in-built system for withstanding pollution hazards.

-- Plants are selective absorbers of metal ions. The mechanism of toxicity is not fully understood. Some concentrate metals in their cell walls without any ill effect. In most cases matals disrupt active metabolic sites within the cells.

-- Plants have ability to absorb P.A.H. (Polycyclic Aromatic Compound) from air water and soil by roots and foliage. Whereas ordinary phenol is not found in plants in a free state, a wide range of phenol compounds, including Catechol, resorcinol and hydroquinone, are generally present. Phenolic compound in the protective covers of plants play a role in disease resistance. Vascular plants have at ability to remove phenolic compounds from polluted water. The uptake of heavy metals by merine plankton is significant.

-- Although most plants die in the presence of 0.1 per cent of Copper in soil, it is remarkable that certain plants can tolerate such conditions, Some grasses have high degree of tolerence. This tolerance is confined to only one metal by a species.

-- Micro-organism (algae, fungi, bacteria and protozoa) are involved in many Key ecological processes such as primary production of organic food materials, and in the cycling of elements. Some microbes are more resistant to metal pollution than higher plants and animals and could, therefore, play an important role on the recovery of polluted area.

Amelioration of tribal belt will need rehabilitation, restoration and revegetation in mine areas :

Mine lands are a fascinating challenge to the foresters. Because the pre-existing ecosystems are extinguished, it's also a challenge to the biologists and engineers to replace them as they were. Some experts are of opinion that no adverse environmental impact should last longer than half a childhood, i.e. 5 years. The restoration of landscape is generally accepted in integral part of mining operation where one have to develop the means of economically establishing permanent, virtually maintenance free, vegetation cover or sub-marginal waste, Ore-dump and tailing. But to achieve permanent revegetation of a toxic, sterile and infertile site we have to synthesise an Ecosystem.

Revegetation entails the establishment not only of plant species, but of biological processes, such as nitrogen fixation and mineral turnover by decomposer organism, site evaluation etc. In gully control, a bag of fertilizer is more effective than a bag of cement.

In planting choice of species the help of geo-botanical indicator (pollution scavenger) may be necessary. Adverse effects of toxic metals are reflected in the morphology of plants which may be a guide to assessing status of soil and plants inter-actions. Lichens have been recognised as an effective pollution monitors and their analysis may indicate the status of pollution in atmosphere. Metal toxicity and their effect on various plants parts have been established and such data may help in assessing the environmental status.

The chemical aspects of soil is a complex study, but such study should be considered essential for any useful planning or revegetation of mining areas.

Initial work of back filling of void, regarding the overburden, proper utilisation of top soil etc. would be necessary. Besides several other technical work should be considered necessary before undertaking vegetation work which are – diversion of surface water from the mines and tailing dam area, drainage stabilisation, collection of all acid effluent to a single treatment either by oxidation, neutralisation and precipitation, disposal of precipitated sludge and the diversion of treated water into the natural stream below the mine working, or by impounding and natural evaporation.

About plants, it has been found that some plants have ability to flourish in elevated levels of trace elements (cryptogams have a high level of tolerance). The tolerance of plant to pollutants can to a certain extent be related to their genetic variability. Some plants have developed inbuilt systems for withstanding pollution hazards.

All these reveal the complexity of the subject.

Social Forestry – to formulate a Strategy

Various environmental, social and economic problems and hazards suffered by the tribals have been enumerated in broad outlines in the foregoing paragraphs. It is obvious, the most serious of the hazards are faced by those tribals residing in the central and western India.

The author considers that in present economic situation of the country only the forestry sector can bring about miraculous and most effective changes in the habitat and ameliorate the hazards of day to day life of the tribal communities working in more than 500 mines in the country.

But the job will be extremely challenging as the tribals are sensitive to any drastic change in the habitat, the soil is sterile and the task of revegetation is fought with various problems. In selecting species various factors enumerated about soil-plant interactions have to be considered.

Social forests will have to devise planting techniques and choose proper species to check erosion in quarry site, over burden dump, spoil dump, abandoned quarry, tailing dams and revegetate all such areas. For improving quality of life there will be need for planting flowering trees, fruits trees, create horticultural garden, mini zoo, demarcate nature reserve and plant various species to supply M.F.P.

Suitable species have to be selected to build-up sound and dust barriers. Proper use of scientific methodology in creating shelter belts in all vulnerable areas have to be made. Various low cost soil conservation engineering designs in controlling gully, sheet erosion, controlling nullah river banks have to be made. Intensive and extensive planting of grass, legumes and shrubs will be necessary.

A very well knitted extension service facilities have to be organised to contact various groups of tribal people to educate them about the hazards they are suffering from and how for tree planting can improve the situation.

Present System of Revegetation in Mining Areas

The present system of revegetation work is done by the mining authorities themselves. They get management plan prepared by experts and competent authorities approve of the plan.

But the author considers this work extremely difficult. Afforestation work cannot be successfully done by the mining authority and only the experts in Forest Department well-supported by research base can tackle such serious problems and Social Forestry wing is the best organization to tackle the situation of revegetation and motivate the tribals for success of the programme.

The finance should come from Mines and Minerals Department of each State, each mine owner contributing money for his own mine, and the Social Forestry Wing will spend the money.

To execute this work with success integrated management plan has to be done over a cluster of mines in an area besides management plan for each mine separately.

Various states have already formed state and district level co-ordination committees on Social Forestry matters. Panchayats have also been involved in various activities. Extension staff have to be carefully selected. Some states have already started establishing Biomass garden, Clonal Bank/Progeny Orchard, Agroforestry, Socio-economic Services, linkage with various rural extension services.

Finally the author wants to stress that the tribal people are to be educated to realise their position and develop desire to have a better quality of life as meant in the following :

A Chinese adage

- If your plan for one year plant rice
- If your plan for ten years plant trees
- If your plan for 100 years educate mankind.

In words of Swami Vivekananda –

"All the wealth of the world cannot help one little Indian Village if the people are not taught to help themselves".

10

MAN, MAN-EATERS AND ROGUES

CASE STUDY

Allow me to narrate today some of the stories about rogue elephants and man-eating tigers, the two most magnificent and popular animals of India, I have collected over the years as conservator of Forests. Some of the stories were presented in International symposiums on Tiger (in 1979) and Elephants (1982). Apart from telling the stories I have also tried to pin-point some of the behavioural characters of rogue elephants and man-eaters giving reasons for those types of behaviour. In fact there are few equals of these two animals in the whole of animal kingdom.

Indian Elephants and Rogues

Indian elephant is a noble animal that has carried oriental potentials on it's flattened back in state ceremonies. No animal has served men more faithfully through the ages and has been made a subject matter of myth and fanciful stories. It is still a puzzle how man failed to understand this animal which lies at the very foundation of the country's symbol of social life, religious ceremonies and in fact a part of his life. Docile nature, strength, courage, endurance and freedom from vices are appellations attributed to Indian elephant.

Yet it has become occasionally insane, determined killer, marauder, vindictive, monotonous killer, revengeful and is prone to make unprovoked attack with evil design and, therefore, the appellation of rogue/killer seems to justify it's habit, pattern. Killer rogues are normally solitary bulls, either tuskers or tuskless, commonly called 'makna'. Cow elephants also turn killer.

Every year at least 40 persons, adding the unrecorded ones it may be 80, are killed by elephants in the country's five scattered zones (north east India, north western Uttar Pradesh, Orissa, hilly areas adjoining the common boundaries of Kerala, Tamil Nadu and Karnataka). From about 150 of such killing episodes the author has classified them into a few types which are presented in short outlines as follows :-

Type 'A' This type of killer chase men and kill the victim by striking with the tusk. If per chance the target is missed the killer does not wait to strike for causing further harm. Such attacks are unprovoked. The killer does not repeat such act of vandalism.

Type 'B' The second type of killer elephant comprise those that are vindictive enough to kill by trampling under foot. The dead body is then lacerated by repeated goring and trampling. The body is then mangled into pulp. Such acts of violence are repeated and more persons are killed.

Type 'C' The third type of killers are craftier that come stealthily and attack the victim when they are most careless. They often take vantage position and suddenly attack the victim with combined force of feet, trunk and tusk. The body of the victim is, sometimes, severed and thrown away. Such act of violence is often repeated.

Type 'D' The fourth type of killers often catch hold of the victim by their trunks and crash the bodies by trampling. The bodies are then made into pulp and put inside a grave dug by the animal and cover them by earth or dried leaves.

Type 'E' These killers (should they be called killers ?) during their night's stroll destroy roadside trees and labours temporary sheds rather out of curiosity and some people get killed inadvertently. In the villages very often people under influence of liquor come on the way of lone elephants and get killed. Such act of killing is not repeated.

Type 'F' These types of killers kill men intentionally but do not indulge in such acts immediately after. They do so after a gap of several years. It seems they act in a rhythmic or intermittent manner.

Type 'G' These type of killers are generally departmental tame female elephants which are docile, faithful and obedient but suddenly lose balance for a moment (there might be some hidden causes) and kill people or even its mahout (the reason may be due to temporary black out). They then forget such incidents and never kill again.

Type 'H' These killers (should they be called killer rogues ?) are generally cow elephants of herds. Very often some calves in the herd go out of the track and enter into homested land out of curiously or in search of food. The mother elephant follow them to retrieve them; in the process some people coming on their way get killed inadvertently.

The aforesaid classification has been based on the results of analysis of many true incidents. Some of the elephants were really killers, the others were normal ones turning killers very temporarily and very few were really rogues; people often got killed because of their own faults or from inadvertent indirect hit. Temporary 'black out' or loss of balance and intermittent killer instinct have been the causes for the kill. It is interesting to mention that most of the killer rogues had a number of morphological defects, such as unequal length of tusks, tusks at different level, injury on the body and on foot, festering sore due to gun shot, magoot ridden sores etc. Psychological loss of balance occurs due to 'musth'. There are some deep rooted causes (injury caused to the animals by some persons years back) which turn some elephant temporary killers.

A reasonable assessment of the behaviour pattern of the killers reveals that there are no inherent or habitual rogues; about 22 per cent of the so called killers are due to loss of temporary balance, another 25 per cent of the killers have been turned so due to deformity of body, about 10 per cent are killers due to adventurous nature, 8 per cent are habitual killers and the balance 36 per cent are circumstantial killers. Of the total 14000 elephants that India has today, only 2.25 per cent could be termed as rogues or really killer elephants.

Man-Eating Tigers

Though we have lot of stories about the cattle lifting tigers few are on the man-eaters of Sunderbans. Jim Corbett's famous stories of man-eaters of Kumaon and Almora hills are legends; of late the man-eaters of Lakhimpur Kheri of the terai region of Uttar Pradesh have created sensation beating, perhaps the guile and ferocity of the Kumaon tigers. Sunderban tigers are not cattle lifters. They are prone to kill men than cattle. Though it is a subject of research it may be emphatically stated that various ecological factors of the estuarine tract have contributed to the peculiar killing syndrome of the tigers where they are supreme in their aggression, tenacity, strength and intelligence. They have a diabolical understanding of human mind, acute power to manoeuvre and are a relentless fighter.

The man-eaters on Sunderbans have taken a toll of 600 lives between 1964 and 1979. Unreported cases may inflate this figure to 1200. The victims are poor villagers whose sheer poor economic condition compel them to brave the hazards of the inhospitable terrain. They face tragic ends; but those return home alive rarely describe the man-eaters in inimical terms. The tigers are also worshipped as 'Southern Lord' or 'Dakshin Roy'; very often the priests put a taboo on the villagers to talk of tigers to outsiders. We are handicapped for want of positive information to bring into light facts about the tigers and men, for these poor people suffer immense molestation from these ruthless maraunders. These supernatural animals operate in a land of sinister beauty, swift flowing water, where armoured deadlines of man-eating crocodiles and eternal silence prevail.

Analysis of over three hundred true stories of death has prompted the writer of this article to classify the behaviour pattern of the man-eaters into six road types which are–

Type 'A' On a bright morning of April, 15, 1969 a full grown tiger was crossing a big river. It's nostrils were flatting above the water. The crew of the launch spotted the animal from a distance as the launch was cruising by. The crew steered the steel hull launch towards the animal. All attempts to shoot and kill the tiger by rifle shots from a rocking launch proved abortive. Further attempts to pierce the animal with bamboo poles also failed. In the process the tiger was protecting itself, while swimming, by tactical manoeuvering for over two hours. As the battle of nerves was reaching climax the tiger astonishingly survived all attempts to suffocate it by driving the launch over it. Suddenly the tiger got air borne and in a superb act of somersault and acrobatic feat landed on the dingi tied at the toe of the launch. After a while the rushing wages overturned the dingi and the tiger thrown into water. The tiger leisurely swam to the shore, walked nonchalantly into the deep forest.

These tigers are superb long distant swimmers; they cross the rivers; they cross the rivers across the current once they have decided to cross a river at a particular point they rush against the stream sweeping aside all nature of obstacle; in water they are as strong as on land; they are extremely crafty fighter with indefatigable energy.

Type 'B' The story narrated hereinafter relates to a fisherman who along with some of his associates was catching crabs on December 26, 1971 at about 4.30 p.m. High tide was fast flooding the slopes of the creek. Yet the fisherman decided to continue crabbing. Unfortunately they were blissfully unaware that a full grown tiger was watching and stalking them. Although they were careful enough to keep the boat in the middle of the stream, they in their preoccupation overlooked the fact that the channel further down narrowed to a bare three metres rendering the boat vulnerable to a tiger's leap from the bank. After the catch the boat made a move to drop anchor for the night. The man-eater anticipated the move and marched off in advanced to take a position behind a keora tree at the narrowest point. Inevitable happened as the tiger caught hold of the boatman with a mighty jump and neatly carried him deep into the forests. The dead body was recovered but the jugular vein and cerebellum were tethered.

These tigers appear all on a sudden in most unexpected place and moment; they attack in broad daylight aiming mostly at vulnerable points which reveal their thorough knowledge of human anatomy.

Type 'C' This story bears witness to tremendous capacity of the tiger to jump into the boat without slight movement, strength and ferocity besides it's man-eating urge.

It happened at 11 p.m. on August 25, 1966. The story goes that Gazel Gazi suddenly awake from deep slumber and found that a pair of piercing eyes was glowing in the poor flickering light of the kerosene lamp in the boat. A warm putrid breath was blowing over his face. The animal which climbed stealthily into the boat anchored in the midstream stood towering over the boat crew to choose his prey Gazi tried to shout, but his voice failed. The man-eater selected Subhas Mandal of 40 years age, lifted him in his jaw and jumped out of the boat and vanished in the darkness.

These types of tiger climb a boat without making any sound, not even a splash is heard. They operate in the morning between 7 a.m. and 9 a.m., in the evening between 3 p.m. and 5 p.m. and between 10 p.m. and 11 p.m.

Type 'D' It was 10 a.m. on April 10, 1970 Ram Barui, aged 50, got down to an island for honey collection. The medicine man (locally called 'Gunin') who, it is said, has power to enchant tiger, accompanying the party had a premonition of grave dangers and ordered that all of them to stay together for some time. His lunch paid off as they sighted a huge tiger. They then returned to the boat. The next morning they stopped out on the dry shore of a creek. Blessings were invoked from the tiger God. Incarnations were chanted to keep the man-eaters away and immobilize them. As the party moved along, Barui for some reasons dropped behind the party. A huge tiger appeared from no where. Barui sighted it and turning towards it shouted while beating the bush with a stick he had with him. But as the tiger was hungry, it snarled, twitched, it's tail and facial muscles expanded, it spit and grazed its hind leg on the ground and visibly grew double its size. Though death was imminent, Barui held the stick high and struck the tiger. This caused the animal to retreat slightly. Barui now touched the ground with both hands and chanted a mantra. The animal probably scent danger and retreated. Barui was rescued, but he ran high temperature; his eyes turned red and remained red till to day.

Thus tigers when challenged pause and stand still; where they seen danger they flee. Are they cowards ? They always attack from behind.

Type 'E' This story refers to Pratap Bauli and three of his men. It was 4 p.m. on April 18, 1970 Bauli was leading his men in forest and was talking while walking. Suddenly he looked back, but could not find his companion; he saw a man-eater hidden behind a bush with one of his men in his jaw till all signs of life vanished while other men were lying unconscious on the ground. Intense love and loyalty prevailed over fear and nervousness. Bauli held a stick in hand and chased the tiger as he shouted. Ram Hawly, however, got entangled in a bush and the tiger could not extricate it. Bauli could hold of a leg and a big tug of war ensued. Fortunately, in the mean time two of his men regained consciousness and joined Bauli in beating the tiger away.

Men are killed by tiger: they lose their near and dear onces to the tigers; every time they go back home, they promise not to come back again, yet they come; this saline tract attracts them; all soft feelings and fear vanish; the tract makes a man tough and face animals without fear and fight for survival.

Type 'F' The story of Md. Sheik stands a blowing monument of the man-eater supernatural and invisible (?) forms. It was 9 a.m. of 5th November, 1970 he had a terrifying experience of life and death struggle with man-eater in the forest. Two watchers were engaged by Sheik to keep a watch over the workers of Golpata coupe. Suddenly it was observed that one of the watchers was missing. A search revealed that there was fresh pug mark of a tiger and a few drops of blood. The medicine man in the party led the search party. He followed the blood trail. On the medicine man was to recover the dead body which was detected. The man

> approached while others were waiting at some distance. From nowhere the man-eater came roaring and savagely pounced on the medicine man from his right, and in full view of all men, carried him away as if to mock at the efficacy of mantras.

The tigers are invisible and appear to be supernatural. They approach with supreme stalth even in broad daylight. It is believed that the medicine man invoke blessing from 'Tiger God' 'Banbibi' or 'Narayani Ma' and can immobilize the man-eaters.

Such are the true stories of the rogues and man-eaters. These true stories show that truth is stranger than fiction. While precaution can be taken in case of rogue elephants, it is difficult in case of man-eaters. The writer has established in his various published papers that the man-eaters operate mostly in the morning, evening and at 11 p.m. at night. They select the victim within a range of age between 35 to 45. Only about 28 per cent of dead bodies are recovered. The people of Sunderbans have accepted the supremacy of the man-eaters and are callous about their safety. But the forest managers are ever conscious of the value of human life and are taking measures to reduce such loss of life. The animals have right to live so also the men. Let of them co-exist.

11

ECO-BIOLOGICAL APPROACH TO SITE SELECTION

Use of modern techniques of site identification for the purposes such as afforestation, pasture development, social forestry site identification, mass tree planting, fodder development, waste land rehabitation have been the cry of the day. Such approaches have been though the use of maps of various scales, preferably of 1 : 50,000 or 1 : 25,000 or even 1 : 15,000, analysis of soil and classification of terrain and through other such conventional methods. Recently aerial photography and remote sensing methods have been accepted as suitable guide to such selection. These methods have been found to be not only quick but also very economical as compared to conventional field surveys for such purposes. Some soil Scientists feel that forest soil–vegetation survey envisages to provide an inventory of the soils occurring in different parts of the forests through survey, mapping, classification and correlation. The question is, how much the satellite imagery can provide. The features that are distinguishable on the imagery (on Bands 5 and 7 or colour composites) during visual interpretation are as under :

(*i*) Land forms or geomorphological features including general topography

(*ii*) Flood plains

(*iii*) Present landuse–degraded forests

(*iv*) Agricultural Crop

(*v*) Field sizes

(*vi*) Linear features like rivers, canals, streams, torrents etc.

(*vii*) Urban Congremerates

Analysis of landsat imagery, computer compatible tape (CCT's), Aerial photography and the use of Electronic Computers are very useful guides. Data acquisition by Earth Resources Satellites is a relatively new technique that is available which can be economic and time effective and help in arriving at a broad land use classification of the geographically area. The data are basically available in two forms such as, an imagery printed on photographic paper and indigitised form on the Computer Compatible Tapes (CCT's). Visual interpretations of transparencies of band 5 and colour composites on a light table can identify broad land-use classes such as–Forest land (Scrub, snow, dense forest etc.), Non-forest Land (agriculture, habitation, water bodies etc.).

National Bureau of Soil Survey and Land-use Planning, Nagpur has used the R.U.B. system in Landsat 3 which contains two broad band T.V Cameras that produce adjacent images. Each Camera responds to rejected energy wave length of 0.5 to 0.75 micrometers and covers a 99 km area with a total swath width of 185 km. Four R.B.V. imagery or sub-scenes coincide approximately with one MSS scene. Each scene has a spatial resolution nearly twice that of the MSS (Watkim, AH, 1978 in the Eros Data Centre Bull).

The author initiated a project viz., "Role of aerial photographs in site identification for social Forestry–a case study in Balaghat District" worked by the officer of Forest Survey of India where Aerial photography (black and white panchromatic on 1 : 50,000 taken by RMK Camera of focal length 15.3 cm) used successfully. In another similar study in Western himalayas Black and White panchromatic aerial photographs of scale 1 : 50,000 was used. Scientists of Indian photointerpretation institute have identified sites for afforestation purposes by using 1 : 10,000 scale panchromatic vertical aerial photographs. Many such examples can be cited.

As such in the present day Site selection for various types of works the use of aerial photographs are being widely used; besides satellite imageries are being used to a limited extend.

These methods would indicate various sites giving them physical characteristics. It defines boundaries of soil units, landforms and land classification boundaries. It is not used, perhaps for identification of exact soil types, but to locate changes in the land-surface pattern that may be related to differing soil properties. The present multi-spectral and multi-band photography is rarely a realistic choice in India for routine soil surveys. Panchromatic photography has been more widely used in soil surveys mapping than colour photography.

The term 'site selection' is generally widely used. The sites may be terrestrial or aquatic. These may concern soil tree, shrubs, herbs, epiphytes relating to various horizontal and vertical strata of vegetation. The sites may be estuarine, pond, marshes or marine. These may also be temperate forests, tropical forests, desert location or mangrove forests. As such, various techniques would be involved to determine what sites as enumerated above would be chosen and for what purpose.

Environmental forestry is a term very widely and loosely used by many. The author would consider in this paper the plants, animals, flowers, fruits, water, soil and all biota and the entire processes–physical, chemical, animate, biological–involved in the ecosystem and the study be made from Botanical, Zoological, Biotic, Habitat, Productivity point of view.

The following studies are essential for ascertaining richness of flora and fauna of a locality.

Site Selection in Himalayan Ecosystem

Generic and specific Diversity Study

The study of number of species in each genus, total number of genera in each family and number of individuals in each species has been found essential to ascertain the status of each species or genus and even the families to ascertain the richness of flora and fauna of a region.

A critical observation on the generic and specific abundance and count of individuals showed that about 50 per cent of the angiospermous families have either 1 or 2 genera and few species among them are gregarious.

In an ecosystem the generic diversity maintains a positive correlation with the specific diversity. In such study, high generic diversity has been found to be associated with high specific diversity value. The study of specific diversity is of prime importance for evaluation of richness of a particular community or ecosystem. A low specific diversity gives a warning signal for existence of a species which may be due to biotic factor, environmental pollution, lack of regeneration and reproductive capacity or over exploitation etc.

It has been found that most of the families having high diversity are fruit bearing and also fodder yielding and obviously attract many species of frugivorous birds (Hill mynah, Horn bills, Barbets, Parakeets, Pigeon etc.) and graminivorous animals (Bison, Rhinoceros, Pigs, Thar, Serow, various deer, Sambar, Nilgai, Wild Buffalo).

In the foot hill and adjacent areas there is heavier concentration of birds and animals which include squirrel, civet, mongoose, weasel, martin, monkey etc.

Observations in the foot hills regions (200 to 1400 metres) show that maximum concentration of birds (Babblers 92 per cent, Laughing thrushes 82 per cent, Warblers 52 per cent, Red start 37per cent, Wood peckers 100 per cent, and similar high percentage of Hornbill, King fisher, Fairy blue bird, Myna and lesser percentage of Barbet, Minivet, Cuckoo shrike, Bulbul, Nuthath, Sunbird etc.) and 80 per cent of animals species (big and small) occurs in this region and in wet mixed forests.

But diversity alone cannot be responsible for abundance of animal. Higher productivity (in dry weight) has been found correlated with animal species diversity. High productivity is derived from stem, leaves, flowers and fruits. But caloric concentration which is of prime importance for animal species may or may not go parallel with high productivity. Caloric production is directly involved in animal nutrition than in dry weight production.

The floral diversity parallel the mammalian and other macro and micro-faunal diversity. Quantitative estimation of flora and fauna of the area indicates this parallelism. The study of flowering and fruiting pattern, type of leaves, their basal area and size classes of fruits (phenophase, phenodynamic and phenometry) testify to the diversity and richness of flora. This diverse flora also maintains a diverse avifauna. As such, diversity study is essential for selection of site for environmental forestry.

Eco-Phenological Studies

The following studies were made which are also necessary to ascertain the richness of the sites for their selection for environmental forests in Moist Deciduous, Semi-evergreen and

Temperate Forests in Darjeeling and Sikkim Himalayas.

- Leaf Surface–800 sq. mm is found in 33 per cent, of species, 801-1800 sq mm in 26 per cent of species, 180-2500 sq mm in 36 per cent of species, over 2500 sq mm in 1 per cent species.
- Types of leaves–Simple alternate forms 51.9 per cent, Compound alternate forms 27.5 per cent, Simple opposite forms 13.0 per cent and compound opposite forms 7.6 per cent.
- Concentration of varied plant forms (per ha.)
- Climber concentration is 285 Nos. Loranthus concentration is 100 Nos. Epiphytes concentration is 285 Nos. Hard fungi concentration is 100 Nos. *Asplenium nidus* with the aid of special overlapping leaves a specialised (biotope) 'niche' that collect little and detritus. Thus epiphytic humus may be even a tonne per ha. which creates new biotope above ground level.
- Trees per hectare–The density is 200-250 Nos.
- Canopy layer–Top canopy has 25.8 per cent of species, middle 48.4 per cent of species and lower 25.8 per cent of species.
- Flowering–Those of monocot families the flowering season spreads over from April to August, in Dieot, Polypetale in April and May, gamopetalae from April to October.
- Fruiting–Fruit of size below 0.5 cm 60 per cent, 0.5 to 1.00 cm-28 per cent, 1.0 cm to 2.0 cm 8 per cent, over 2 cm-4 per cent. The size of fruit has positive correlation with the size of frugivorous birds. Ficus sp. have a density of 2-3 tree per ha. Some of the species throw 500-1000 kg of fruits a year and they are great attractions for various species of birds.
- Insect and Colouration of flower–Population and distribution of some groups of insects are to some extent dependent on colouration and extent of flowering. An analysis of colouration shows 28.5 per cent is white, 24 per cent is yellow, 10.4 per cent is pink, 12 per cent red and 25.5 per cent is of other colours. Various grades of cent also attracts wide groups of insects.
- Fodder and dietary need of animals–Eastern India is very rich in the fodder resources. High incidence of occurrence of species per ha. and wide varieties of grass, herbs and shrubs

have enriched the fodder resources of the wildlife as well as of domestic animals.

Ectone Zone

Champion and Seth have identified more than 120 forests types in the forests of Eastern India. This complex, habital well indicates the occurrence of innumerable ecotone zones to attract various species of animals. Such zones are essential for diversity in plant and animal species.

Studies on Plant-animal interaction

Plants and animals have intimate reciprocal relation and are adopted to the local environmental condition. The physical and chemical factors in the habitat are responsible for the size and composition of the communities.

Animals depend on plants, either directly or indirectly for food. Insect, bird and bat visit flowers to feed on the nectar and pollen. Scent and colour of flowers and nutritive value of nectar or pollen–the biochemical factors are associated in this interrelationship. Pollination biology is a complex process of interaction between plant and animal. Ruthless exploitation of forests may bring a drastic change in such interaction.

Flower colour, shape of flower, scent, etc. are responsible for various species of insects and birds being attracted towards them. Bees are sensitive to red, yellow and blue colours. Hive-bee prefers blue and yellow flower, bats prefer white or drab green colour, beatles prefer dull cream or greenish colour. Moths prefer red, purple or green colour and Wasps prefer brownish flowers. Birds prefer scarlet, yellow or red flowers. In nature colouration of flower in order of abundance is white followed by yellow, red, pink, blue, etc.

Insect behaviour is known to be controlled by chemical signals, which have the form of volatile organic constituents released by one insect to affect another. These are termed "pheromones" (Hormones of higher animals). Pheromones are involved in almost every aspect of insect life feeding, sex, aggregation, oviposition, defence and laying trails.

These insects from prey animals to visit number of birds, lizards and amphibians.

The nature, as such, is maintaining a perfect balance in various spheres of plants and animal life and there is immediate necessity to select such sites and conserve in entirety.

Distribution and Concentration of Insects

The distribution of several insect orders show abundance of Coleoptera between 1250 and 1800 metres, Dermaptera between 1250 metres and 2250 Hymenoptera between 1300 and 2200 metres, Diptera between 1500 and 2200 metres and Lepidoptera between 900 and 2200 metres. Coleoptera and Lepidoptera are two insect orders mostly concerned with plants and they surpass numerically all other insects in their number and distribution. Size analysis of Coleoptera shows the following :-

2.5 cm about 47 per cent, 2.5 to 5 cm about 32 per cent, 5 to 7.5 cm about 14 per cent and 7.5 cm and above.

Butterflies and Moths are most conspicuous in the forests owing to their elegance, colourful pattern and wide range of damage to various organs of plants. India has about 1443 species of butterflies of which Eastern Himalayan region has 835 species. Moths (Sphingidae) in the Eastern Himalayas have 135 species. Maximum concentration of moth and butterflies in between foot hills and 1200 m. Dinaids are common between 900 and 1200 m. The families *Myrsinae, passifloraceae, Urticaceae, Compositae, Rubiaceae, Moraceae, Ampelidae, Lequminosae, Asclepiadaceae* and *Apocynaceae* are principal host plant families of Moths and Butterflies.

All these show abundance of population of insects of various types at different altitudes and geographical ranges in the lower hills. In the high altitude diverse type of insect communities also occur in barren rocks, on alpine meadows, on snows, on aquatic habitat, on damp moss covered rocksurface, in rock crevices and in soil. In such situation living conditions are extremely difficult, organic and inorganic food limited, sunny hours very restricted. High wind, extreme cold, physiological drought compel the insect to localised heavy crowding in limited spaces (microclimatic niches). Even the high altitudes of Western Himalayas, 2500 species with more than 400 species occurring near tree lines have been recorded. Principal orders to which these insects belong are Coleoptera, Lepidoptera, Diptera, Crinoptera, Odonata, Collembola, Plecoptera besides Scorpions, Mites, Spiders etc. in western found to be 3.34.

A linear Conservation Reserve from foot hills to 3700 m. in West Bengal and from foot hills to 5000 m. in Sikkim in vertical strip with link as far as practicable would from ideal habitat for innumerable species of insects, birds and animals.

Distribution and Concentration of Avifauna

Analysis of various data have revealed the presence of (*i*) 650 species of birds in the Darjeeling hills (*ii*) 400 species of birds from the Bengal Duars forests (*iii*) 237 species from Moraghat forests (*iv*) 75 species from Mongpoo and 121 species from Hantupara alone.

Abundance of Avifauna species have been recorded in the foot hills upto 1000 metres or as high as 1500 metres. This zone correlated directly with the abundance of flowers, fruits and insects. Analysis of data of East Himalayan birds reveal that the zone from plain to 900 m. altitude contains 50 per cent of the avifauna species. Considering all these facts, the foot hill habitat seems not only ideal for most of the avifauna species, but also for scores of flora, insects, mammals and other animals.

Distribution and Concentration of Amphibia and Reptiles

Qualitative assessment about these animals is lacking. The genera *Bufo, Rana* and *Rhacophorous* are three principal genera of amphibia with 125 species that occur between 900 m and 1800 m. The abundance of reptiles has, it seems, direct correlation with the abundance occurrence of amphibia, insects etc. in the Himalayan hills. 72 species occur between 600 and 900 metres. 12 species between 900 and 1500 metres and 24 species between 900 and 1800 metres.

All the diversified species of plants, Reptiles, Amphibia and other animals present an extensive and intensive floral and faunal complex in the Eastern Himalayan hills.

Sites for Environmental Forests

In the foregoing paragraphs the necessity of the study of (*i*) diversity in flora and fauna (*ii*) eco-phenology (*iii*) ecotone zone (*iv*) plant animal interaction (*v*) insect concentration zone (*vi*) Avifauna concentration zone and (*vii*) Amphibia and reptile concentration zone has been stressed and areas identified. The Singalila hills and other temperate hills which are moist, wet mixed forests, ecotone zones of temperate and tropical foot hill forests are very rich in micro and macro fauna. The zone between 1000 m to 1500 m seems to be very suitable for the plains and hills species intermingle are also suitable sites for environmental forestry.

Site Selection in Estuarine Ecosystem

Estuarines are special bio-tope in littoral zone. They are most dynamic, complex and tich ecosystem. Here the opposite currents meet, oscillatory tidal currents are formed and sedimentation system is complicated. Estuaries are complex biological environment which also provide aesthetic enjoyment and recreational sites. They are priceless natural assets and are dynamic ecological system, nursing grounds, sheltering place, source of nutrients for the marine life.

Of all estuaries of the world the "Sunderban" estuary is the most complex-one of the largest and has vast assemblages of vertebrate and invertebrates with Royal Bengal tiger at the apex of animal pyramid.

Estuaries are Dynamic Sites

In this ecosystem a scientist is to determine :

(*i*) Salinity in soil and its effect on plants and animals.

(*ii*) Salinity in water and its effect on plants and animals.

(*iii*) P.H. value of soil and its effect on plants and animals.

(*iv*) Various mineral nutrients in water and their effect on plants and animals.

(*v*) Oscillating water currents and their effect on plants and animals.

(*vi*) Various temperature range in water and its effect on plants and animals.

(*vii*) Light and its effect on plants and animals in water.

(*viii*) Moon and its effect on plants and animals behaviour and reproduction.

(*ix*) Tide and its effect on plants and animals behaviour.

(*x*) Temperature and its effect on plants and animals.

(*xi*) Salinity and osmoregulation.

(*xii*) Wind velocity and its effect on plants and animals.

(*xiii*) Sandy clay and rocky shore and animal dispersion.

(*xiv*) Inter tidal mudflat and animals.

(*xv*) Effect on salinity and other factors on the behaviour of animals and especially on tiger.

(*xvi*) Density and pressure of water and its effect on plants and animals.

(*xvii*) The cause and effect of bioluminesoence.

(*xviii*) Various plant and animal zones.

(*xix*) Wide fluctuations of tidal level.

Salinity of water and soil is of prime consideration as on it depends the variety of plants and their physical, physiological and chemical adjustments. Not only the plant and animal species are specific and limited, animals in such areas have to develop various physiological adjustment. A special chemical environment is produced owing mixing up of various concentration of salts. In summer salinity increases. There is steep vertical salinity gradient (high at surface and low towards bottom).

Endomosis and Exosmosis must be regulated (camoregulation by all categories of animals and plants). Animals response to sudden changes in salinity and the following :

(*i*) Immediate death.

(*ii*) Autonomy (self amputation of a part as in certain worms, anthropoids and lizards).

(*iii*) Violent contribution or irregular movement.

(*iv*) Tropism.

(*v*) Abnormal physiological function.

(*vi*) Establishment of a balance

(*vii*) Acclimatization of the surroundings.

Ecological Biochemistry in Estuaries

A new interdisciplinary subject–ecological biochemistry or sometimes named as chemical ecology or phyto-chemical ecology–

is of prime concern. More and more chemical techniques are being used in biological systems. It is also due to awareness of ecologists that chemical substances and particularly secondary metabolites such as alkaloids, flavonoids and terpennids have a significant role in the complex interactions occurring between animal, animal and plant or plant and plant in the natural environment.

Ecotone Region–Estuarine/Marine Region

Geographic distribution of marine organism depends on their responses to current, temperature and physical barriers, their local distribution is affected by waves and tides, type of bottom salinity and depth.

Major communities recognised are the oceanic plankton and nekton biome in the open sea, the balanoid gastropod thallopsite biome on rocky shore, the pelecypod–annelid biome or sand and mud bottom and the coral reef biome, coral reef are found only in the tropics. The warm water faunas are richest in species specially in the Indo-Malaysian and West Indian sub-regions.

Marine plankton includes a greater variety of forms than does fresh water plankton. Abundance of plankton varies from place to place and from season to season. Plankton is much more numerous in coastal water than in open sea. Diel movement between the surface at night and greater depths during the day are pronounced.

Benthos (bottom) includes a great variety of sessile creeping and burrowing forms. It is very abundant in littoral zone and decreases as depth increases. There is considerable difference in life form and muddy ones. Zonation of species is most prominant in rocky than in depositing shores.

Diurnal Changes

Diurnal changes in the movement of planktonic animal which rise to the surface of water at night and by day move into deep zones.

Currents and Oxygen

Current of air and water are predominant feature of many habitats and they are among the environmental factors that determine the distribution of animals.

Daily occurrence of high and low tide twice and also to innumerable creeks and rivers anastomising horizontally and vertically cause various complicated reversing currents. The animals and plant organism have to develop special modifications to adjust to such environment.

Influence of Moon

The moon is associated with fertility reproductive processes and growth. More of fish catch is reported immediately before full and new moon. Endogenous and exogenous systems in animal are concerned in various physiological activities.

Tidal and lunar fluctuations exert influence on organisms inhabiting the tidal zones. They may be subject to daily rhythm, seasonal cycle or in coastal areas to the turns of the tide and the phases of the moon. Internal physiological changes (endogenous) in most cases and not external circumstance (exogenous) are related to the change. Endogenous Self Sustaining Oscillation (ESSO) is term used to explain the phenomenon (Example-Uea sp).

Sites for Environmental Forestry in Estuaries

The dynamic estuarine regions differ widely among themselves. Of all estuaries of the world the 'Sundarbans' stand supreme in its peculiar type of estuary. Wide difference in low and high tide level, amphibian, Royal Bengal tigers which are man-eaters and having an assemblage of wide variety of animals (vertebrate and invertebrate) make the area a distinct type. It is possible to select specific areas having diverse plants and animals species and having wide characteristics. Such estuaries are facing destruction and pollution owing to several biotic factors. In order to solve problem involving physical, chemical and animate environment in the estuary multidisciplinary approach is imperative.

SECTION B

INDIA SAVES ITS VARIED WILDLIFE ECOSYSTEM

12

WILDLIFE ECOSYSTEM IN INDIA – INDIA SAVES ITS NATURAL RESOURCES –

India's rich, magnificent and varied wildlife in diverse zoo-geographic, eco-climatic and phytogeographic zones has been her traditional heritage and priceless assets. The beauty and variety of wildlife resources have no limit. More than 500 species of mammals, 1200 species of avi-fauna and 30,000 species of insects, 900 species of reptiles and myriads of other living organisms stand unsurpassed in her faunal resources matched by her equally diverse vegetation types.

The animals include magnificent gigantic elephants associated with the splendours of our princely pegeantry, the largest bovine such as gaur or Indian bison, the one-horned rhinoceros considered the greatest of all rhinoceros, the swamp deer, the breath-taking spotted deer and lastly the picturesque nilgai and the like. There are no parallel to stately lion, the magnificent tiger, the splendourous clouded leopard, the dazzlingly painted vermillion bats. Of these highly diversified and colourful faunas in India today, some however, are on the verge of extinction. Unfortunately these fauna are exhibiting retrogressive tendencies owing to biotic interference.

Unique Biogeographic Realm

India has a unique bio-geographic position which is in the confluence of palearctic Ethiopean, Indo-Malayan and Indo-Chinese realms of fauna and possess a number of components from each of these realms. The Ethiopean region is represented by chinkara, the hyena and the ratel; the palearctic by lynx, wolf, markhor, urial and hangul; the Indo-Chinese by red-panda and the mask deer; the Indo-Malayan by the hoolock and gibbon, the elephant, the goat antelope, and mousedeer.

India along with south-eastern Asia forms a part of the oriental region. This is one of the six great geographical realms into which the earth is divided. Each of these realms supports its characteristic animal life. The Indian region has varying physical characteristics along with a great variation in climate and vegetation. Zoo-geographically India can be divided into several zones which are : (*i*) Malabar coast, (*ii*) Main peninsula, (*iii*) Gangetic plain, (*iv*) Indian desert, (*v*) Deccan thorn desert, (*vi*) Punjab and its semi-arid desert, (*vii*) Ladakh and its cold desert, (*viii*) Assam and contiguous hills, (*x*) Wooded hill slopes of the Himalayas (*x*) Andaman and Nicobar islands, etc.

The distinct zones are indicative of the richness of fauna of the Indian sub-continent from ecological and bio-ecological points of view. India's unique diversity in fauna stems from her ecological variation typified by hot rainless deserts, cold snowbound mountain peaks and wet and soggy rainy regions.

National Wildlife Action Plan

Government of India has in recent past (1981) brought the Department of Environment, Forests and Wildlife under one Ministry to facilitate proper coordination and integration of these related disciplines. An institute named as "The Wildlife Institute of India" for imparting "Training education and carrying out research on Wildlife". In 1952 'The Indian Board for Wildlife' was established. In 1982 'The Wildlife (Protection) Act was enacted to ensure stricter protection of Wildlife and its better Management. The Forest (Conservation) Act was passed in 1980 to impose a severe restriction to diversion of forest land to non-forest use. A National Wildlife Action Plan has been launched in November 1983. The centre provided 100 per cent financial assistance to the states and Union territories for instrumentation of approved item of work relating to the national parks and tiger reserves. New Schemes have been formulated for captive breeding and for rehabilitation of endangered species.

The National Wildlife action plan has the following salient features :

(*i*) Establishing a net work of scientifically protected areas such as national parks, sanctuaries and biosphere reserves to cover adequate geographical areas.

(*ii*) Developing appropriate systems of management concerning protected areas and restoration of the degraded areas.

(*iii*) Building up a professional cadre of personnel fully trained for the management of sanctuaries.

(*iv*) Providing corridors linking all the sanctuaries of a particular area of zone.

(*v*) Rehabilitating indigenous, endangered and threatened species of fauna and flora.

(*vi*) Supporting the management of captive breeding programme for plants and animals.

(*vii*) Developing research and monitoring facilities.

(*viii*) Promoting support for wildlife education and research.

(*ix*) Revising statutory provisions providing protection to wildlife and regulating all forms of trade in wildlife.

(*x*) Enlisting support from and collaborating with voluntary bodies and agencies in the total wildlife conservation effort.

Threatened and Endangered Species

This article is not the proper place to give a comprehensive list of the threatened species of fauna and avifauna. Several species have been extinct and many others are threatened with extinction. The hunting leopard, the two horned rhinoceros and the pink headed duck are few amongst others which have been extinct. At least seventy one species of mammals, seventeen species of amphibians and reptiles and forty seven species of birds face the threat of extinction. Apart from the vertebrates, crustaceans are also on the point of extinction and special mention in this connection may be made of the robber crab of south sentinel island.

The time is ripe for the congregationalists to devote more attention to keep alive the following species to preserve India's rich faunal diversity and take remedial measures :

Mammals–Binturong, black buck, browantelered deer or thamin, caracal, clouded leopard, Tree shrew, Phayre's leaf monkey, Markhor, Pallas, cat, Rusty spotted cat, Tibetan antilope, Tibetan gazelle, desert cat, dungong, fishing cat, golden cat, golden langur, Himalayan thar, tailed macaque, loris, marble cat, mouse deer, muck deer, pigmy hog, one-horned rhinoceros, snow leopard, swamp deer, urial, wild buffalo and others.

Amphibians and reptiles–Gharial, mugger, salt-water crocodile, agra-monitor lizard, green sea turtle, hawk-bill turtle, salamander, Indian egg eating snake, leathery turtle, logger-head turtle, olive black logger head turtle, viviparous toad, water monitor and many other species.

Birds–Andaman teal, Bengal florican, black-necked crain, blood pheasant, narcondom hornbill, nicobar megapode, Nicobar pigeon, monal pheasant, lammergeur, great Indian bustard, spotted owlet, jerdon's courser, Crustacea–Robber carb.

Endemic Animals

The endemic species are not safe as they are confined to a very restricted habitats with long range of distribution and comparatively safer. Many species which were once wide spread have now been driven to a very limited area owing to destruction of habitats.

Species now confined to small area–March crocodile, gharial, One horned rhinoceros, swamp deer black buck, the asistic lion, Royal Bengal tiger, white winged wood duck, the mountain quail, the great Indian bustard, the Bengal florican, the spotted owlet, the lion tained macaque, red panda, the caracal, the Indian clouded leopard, the Indian tiger, the snow leopard, the one horned rhinoceros, the wild ass, the pigmy hog, the Himalayan musk deer, the black buck, the wild buffalo, the gaur etc.

Endemic species–Narcondom hornbill, nicobar megapode, robber crab, Jerdon's courser, the nilgiri langur, browantlered deer, the hispid hare, the markhor, the nilgai tahr, the wild yak etc.

National Parks, Wildlife Sanctuaries and Other Reserves

India's vast wildlife resources have suffered a wide-scale annihilation during the period from 1940–1970. It goes to the credit of the foresters of India that they have been able to preserve most of its fauna from being extinct in her national parks and wildlife sanctuaries. But time has not yet come for complacency. In spite of several wildlife and game preservation acts having been passed outlawing indiscriminate killing of wildlife, national parks and wildlife sanctuaries have become the main refuge of wildlife.

India should feel satisfied, but not complacent, for having saved its wildlife in entirety including those threatened in its 47 zoological parks, 54 national parks and 372 wildlife sanctuaries covering an area of about 1,00,000 sq. km. (figure upto 31-3-85). (54 National Parks cover an area of 21,003 sq.km or 0.6% of land area, 372 WLS cover an area of 88,649 sq.km or 2.7% giving a total of 426 protected areas with an area of 109,652 sq.km or 3.3% of land area). These parks and sanctuaries are distributed all over the country covering 110 climax types, 53 sub-types, 24 edaphic types, 22 seral types and 9 degraded types. These cover ever-green, semi ever-green, moist deciduous, Hollok, Littoral, Mangroves, Tropical swamp, Tropical thorn, Dry evergreen, Desert grass and scrub, river and several other forests.

Conservation projects for endangered species like hangul (1970), lion (1972), tiger (1973), crocodile (1974) and browantlered deer (1981) are sincere efforts to protect some threatened species.

The national parks have been set aside for the protection and conservation of outstanding natural and plant communities, for geological formations and for scenic beauties for the enjoyment of present and future generations in a manner consistent with scientific, cultural, educational and economic purposes. The wildlife sanctuaries have, however, been set aside for conservation and management of wildlife and the protection and management of habitats.

A Bonanza of Diverse Ecosystems

India has been gifted with myriads of micro and macro eco-types, biotopes and complex of symbiotic plant-animal relationships in several ecosystems some of which are as under :

Aquatic and marsh ecosystem (representative sites)–Kolleru pelicanry, Nelapattu bird sanctuary, Nal-sarovar bird sanctuary (all in Andhra Pradesh), Sultanpur bird sanctuary (in Haryana), Chataprava and Ranganthitoo bird sanctuaries (in Karnataka), Chana bird sanctuary (in Rajasthan), Valangudi (in Tamil Nadu) and several others.

Estuarine ecosystem (special site)–These are an asset to this country life in Sunderbans (West Bengal) and Bhittar-Kanica (Orissa). This ecosystem is the most complex of all ecosystems anywhere in the world.

Lagoon ecosystem–Chilka (Orissa), Point calemere (Tamil Nadu) and Pulicat (Andhra Pradesh) sanctuaries are complex sites for micro and macro plants and animals. Aquatic flora which has become a rare commodity in this country will be saved in these sites.

Volcanic ecosystem–The Barren and Narcondom islands in the Andaman group are of volcanic origin. Narcondom hornbills are endemic to Narcondom island only. These are some feral goats in the Barren island. A thorough study of flora and fauna in these islands will reveal more interesting data.

Desert ecosystem–The Rajasthan desert sanctuary is a unique site for the conservation of desert flora and fauna.

Marine ecosystem – India has established marine natural parks at gulf of kutch, and at Wandoor in Andaman.

Scores of marine animals and plants will be protected in these sanctuaries.

Marsh ecosystem–A grand assemblage of aquatic plants and insects and birds may be found in the sanctuaries at Paknal, Pocharam, Dudhwa, Mahavira, Harina, Vanasthali, Kaziranga, Lakhawa, Orang and in several other such sanctuaries.

Tropical forest ecosystem–The tropical forests of Andaman and Nicobar islands shelter endemic Megapode, nicobar pigeon, Andaman teal, Andaman pig have been given protection in South sentinel, Barren, Button islands and at Saddle peak and at Mount Harriet sanctuaries and national parks.

Sanctuaries in Various Geographical Zones

According to Champion and Seth two major types of vegetation cover about 70 per cent of the forest types of India; these are–tropical moist deciduous (30.9%), tropical dry deciduous (38.7%). Other types are–tropical evergreen (6%), tropical and semi-evergreen (2.5%), littoral and swamp (0.9%), thorn forests (3.6%), dry temperate (0.3%), and others (8.4%). Of the distribution of parks and wildlife sanctuaries, the western ghat areas have 32, the western himalayas 33, the eastern ghats 15, Bhutan and Nepal 10, the eastern himalayas 15, the mangrove sites 3, the Indogangetic plains and foot hills 25, dry deciduous and scrub forest areas 80 and the balance in other areas 20.

Rehabilitation of Endangered Species (some major ones)

Crocodiles and gharial–At least 25 sanctuaries and parks offer

Blyth's Traqopan—A colourful bird of eastern Himalayas

Lesser Panda of eastern Himalayas

Aquatic ecosystem; lake, water plants, birds and small animals

Continental Shelf (Coral) ecosystem

Desert (Rajasthan) ecosystem

protection to salt water crocodile, marsh crocodile and gharial. Massive breeding programmes have undertaken to protect these reptiles–at Manjira, Corinaga, Satkosia, Jawahar Sagar, Krishna, Lanjamagadu, Chambal, Bhagabatpur, Nagarjun Sagar, Guindy, Nandankanan, Katerniaghat, Amrawati Kinnersani, and at several other places.

One horned rhinoceros–This rare animal is being given protection at Kaziranga, Manas, Sonai-Rupa, Orang, Lakhawa, Jaldapara (all in India) and at Chitwan-Rapty, valley (in Nepal).

Tiger–In all the fifteen project tiger areas this animal is said to be breeding to expectation (present population 3000). Sunderban is a unique site where the tigers lead an amphibions life in Salt water. Other project sites are Dudhwa and Corbett Uttar Pradesh, Manas in Assam, Buxa and Sunderban in West Bengal, Palaman in Bihar, Simlipal in Orissa, Siriska and Ranthambhore in Rajasthan, Kanha in Madhya Pradesh, Namphepa in Arunachal Pradesh, Melghat in Maharashtra, Bandipur in Karnataka and several others.

Wild buffalo–This rare animal is receiving attention in Kaziranga, Manas, Lakhawa, Garampani and at Kutru Wildlife Sanctuaries.

Black buck–This rare animal is being given adequate protection in Taroba, Pocharam, Kewak, Bori Valavadar, Gir, Siriska, Ratapani, Panch, Kenawat, Kaimur, Tadwai, Nagarjun Sagar, Tungabhadra, and Kanha national parks and sanctuaries.

Nilgai–This animal is being well looked after at Ranthambhore, Siriska, Yawal, Jaisamad, Nawagaon, Pochamarhi, Porch, Gir, Najira and Gomardha sanctuaries and national parks.

Florican–It has a very limited distribution probably in Orang in Assam only.

Mouse deer–This animal is being given protection at Molem Dalma, Borvili, and Simlipal wildlife sanctuaries.

Wild ass–The only sanctuary where it is getting protection is in Rann of Kutch.

Musk deer–This animal is getting protection at Dachigaon, Rajparian, Chimnai, Rohru, Tundah, Talsa, Kinnarnur and Rajgarh Wildlife Sanctuaries.

Ibex, Bhoral and Markhor–At the sanctuaries at Rohru, Tunda, Kugli, Kelatop and Khajyor this animal is getting protection.

Swamp deer–At Kanha, Kaziranga, Dudhwa Parks this animal has been safely breeding.

Flamingo–This bird is to be seen over a limited area in Chilka and Nalasarovar Wildlife Sanctuaries.

Pelican–This bird is to be seen at Chilka, Kolleru and Nelapattu Wildlife Sanctuaries.

Hangul–This animal is to be seen only at Rajparian sanctuary in Kashmir.

Narcondom hornbill–It is to be seen only at Narcondom island sanctuary.

Rober crab–In the south sentinel island this crustacea can be seen.

Brow antlered deer–Keibul Lamjao National Park in Manipur is the only home of this deer.

Great Indian Bustard–This bird can only be seen at Bustard (in Maharashtra), Desert (at Rajasthan) and at Ranebennur (at Karnataka) wildlife sanctuaries.

Golden langur–Only to be seen at Manas Wildlife Sanctuary.

White tiger–It is no longer can be found in forests.

Outstanding Features of a Few National Parks and Sanctuaries

Of 247 wildlife sanctuaries and 53 national parks few sites have special features as (*i*) Wide coverage of latitudinal, longitudinal and altitudinal ranges, (*ii*) High density of animals species (*iii*) High density of individual species (*iv*) Rich floral species and (*v*) Diversities of sites. These qualities may be found in several parks sanctuaries, but some special sites are :

Kanchanjungha national park in Sikkim–It fulfils the above qualities (*i*), (*iv*) and (*v*). Snow leopard, clouded leopard, binturong, Tibetan wild ass, blue sheep, takin, musk deer, marble cat and myriads of insects, butterflies and moths and birds are special feature of this sanctuary.

Namdhapa wildlife sanctuaries in Arunachal Pradesh also fulfils the condition as the aforesaid one.

Corbett national park–fulfils the aforesaid requirements (*ii*), (*iii*), (*iv*) and (*v*).

Simplipal Tiger reserve fulfils the qualities (*iv*) and (*v*).

Kanha national park fulfils the qualities (*ii*), (*iii*), (*iv*) and (*v*).

Nandadevi, Kedarnath and Valley of flowers sanctuaries–fulfil requirements (*i*), (*iv*) and (*v*).

Sunderbans tiger reserve fulfils the qualities (*ii*) to (*v*).

Kaziranga national park fulfils the requirements (*ii*) to (*v*).

Bandipur-Wynad-Nagarhole and Mudumalai wildlife sanctuary complexes probably one of the best sites fulfilling all the conditions.

Abundance of chital in Palamau, Kanha, Corbett and Bandipur, Sambar at Ranthambhore, Swamp deer at Kaziranga and Dudhwa, Hog deer at Kaziranga and wild buffalo at Kaziranga are some notable features.

Wildlife Research

Research in the field of wildlife is still in the incipient stage. The study of varieties of flora and fauna, their interaction symbiotic relationship, diseases, breeding etc. offer rare opportunities for research. Few countries in the world will be able to offer such diverse opportunities.

Decline in Animal Species and Regression

During the last few centuries the vertebrate fauna has reduced considerably. Their composition was more complex. The intensity of this impoverishment is high. There has been qualitative and quantitative decline in the number of genera and individuals. Of species leading to a high degree of regression. The degree of such regression surpasses that of faunal impoverishment.

Mean faunal regression has been calculated by Dr. A.K. Mukherjee of ZSI in his paper entitled "Some examples of recent Faunal impoverishment and regression" (1974). He has found out that the rate of regression is 90 per cent. He used the formula :

$$\frac{(R-r) \times 100}{R} = Q$$

Where Q = coefficient of faunal regression; R = size of the former distribution range in sq.km; r = present size in sq. km.

Why Conservation Programme ?

Flora and fauna are an integral part of the vast natural resources which are in perfect balance with each other. Efforts are being made to create vast areas of biosphere reserve primarily for the protection of the plant gene resources. The massive human population explosion and technological innovations have upset the balance. In the conservation of wildlife, air, water, soil and living organisms are being conserved. Innumerable plant species are also being protected. The conservation has intrinsic cultural and scientific value. 'UNESCO's Man and Biosphere' programme is concerned with these conservation problem and conservation education. In this country the population pressure and extension of agricultural land for optimal food production have over-shadowed the conservation needs. Ecological principles have not received the attention as they should get, Indiscriminate hunting and food scarcity have intensified the capturing of animals, birds, snakes, lizards, frogs and even insects for the purpose of protein food.

Fish population and other aquatic life have been also depleted owing to shrinkage of aquatic sites. However it is well to mention here that the faunal wealth in the tropical and high humid temperature condition has been responsible for creating very stable system in our country in which climate, soil, vegetation and fauna exist in an extremely complex state of dynamic equilibrium. Wildlife which is the faunal component of the ecosystem is closely associated with vegetation. But monoculture of economically important species tends to destroy the diversity of flora and fauna.

Lack of Strong Policy

There has not been a clear environmental policy in this country. Of late government and non-governmental agencies have been involved in the protection of environment. There has not been adequate public awareness of the importance of plant and animal life, of geological features scenic beauties beating on our national life. People should be made aware of the utility of wildlife reserves which are not luxury but a crying necessity. The introduction of wildlife protection Act of 1972 has been a strong and useful step in this direction. Wildlife acts of 1873, 1878, 1932, 1933 and setting up of Indian Board of Wildlife (1952) are also bold steps in conservation effort.

A Great Stride in Conservation Effort

Of late conservation has become an integral part of all development programme in this country. Ecological consideration are being given adequate attention while undertaking all developmental works. Clear guidelines have been formulated in this regard. All planning and executing agencies have been instructed to preserve all natural resources while developing it. A separate Department of Environment has been set up which controls all environmental problems including the wildlife wing of the forestry sector.

Efforts of the state governments to conserve and rehabilitate various species of wildlife in diverse sites have been commendable. Wildlife protection laws provide severe punishment for the law breaking and poachers. Yet the lure of the rhino horns as aphrodasiac, tiger, leopard and lizard skins for high commercial value, snakes, frog, turtle and birds of protein food has induced a large group of people to take poaching. As such the final battle of conservation of wildlife is yet to be won. The guiding philosophy of development should be 'development without destruction.'

National Park, Wildlife Sanctuaries and Other Protection Network in India

The wildlife institute of India under the supervision of W.A. Rodgers and H.S. Panwar a comprehensive report entitled "Planning a Wildlife Protection Area–Net Work in India" in Volumes I & II. This was prepared on the basic report submitted by twenty four experts deployed for the purpose under guidance the offices of the Wildlife.

The present list produced here-in-after is based on old report prepared by the Government of India. This report has been upgraded based on the Institute report. In the decade from 1981–1990, many sanctuaries have been newly created, many denotified, many upgraded as National parks; as such a report is never final and there is always a gap. In the present report the status of some Wildlife Sanctuaries (WLS) and National Parks (NP) could not be ascertained and, therefore, the list may be considered as a broad classification; area of some could not be correctly ascertained.

The following are a brief Summary of (*i*) Conservation status of ungulate species in India, (*ii*) W.L.Ss. and N.Ps. in various Front types of India and (*iii*) Threatened mammals in various protected areas (*iv*) Existing W.L. Sanctuaries and National Parks of India, (*v*) Proposed protected areas.

(*i*) **Conservation Status of ungulate species**

Endangered	Threatened	Rare	Satisfactory
Argali/Nayan	Bharal	Buffalo,	Barking deer
Himalayan Tahr	Elephant	Chowsinga	Black buck
Hangal	Rhinoceros	Goral	Chital
Markhor		Ibex	Chinkara
Musk deer		mouse deer	Hag deer
Tibetan gazella		Takin	Nilgai
Tibetan antilope		Serow	Wild boar
Wild Ass (Ubelan)			Sambar
Wild Ass			
Swamp deer Dry.			
Swamp deer wet (Vulnerable)			
Sangai			
Pigmy hog			
Nilgiri Tahr.			

Extinct–Shou, Yak.

(*ii*) **Some Important N.Ps. and W.L. Sanctuaries in Various Forest types (Selected)**

	Forest Types		National Park/WLS
1.	West Coast Tropical Evergreen Forests	–	N.Ps. Brahmagiri, Karimpuzha, Agastyamalai.
2.	Assam Diplerocarp Wet Tropical Evergreen	–	Tirap N.P., New WLSs in Assam.
3.	Assam Mixed Tropical Evergreen Forests	–	Balphakran, Pahhuni WLSs.
4.	Andaman Tropical Semi Evergreen Forests	–	Many WLSs created.
5.	West coast Tropical Semi Evergreen Forests	–	Koyna, Kudremukh, Agasthamata, Anamata (New NPS).
6.	Orissa Tropical Semi Evergreen Forests	–	Extended Simlipal, New areas in Eghat.
7.	Andaman Tropical Moist Deciduous Forests	–	Many WLSs created.
8.	South Indian Tropical Moist Deciduous Forests	–	Bhadra NP, Bamragarh NP, Extended Nagarhole, Satpura N.Ps., Purna WLSs.
9.	Low Alluvial Savannah Wood Land	–	Manas, Kaziraya, D'Ering, Dibru.
10.	Littoral Forests	–	Many WLSs in Andaman & Nicobar.
11.	Tropical Seasonal Swamp Forest	–	Manas WLS.
12.	Southern Dry Teak Forest	–	Khalibil NP in M.P., Sitamata NP in Rajasthan.
13.	Southern Dn. Mixed Deciduous	–	Nagzira, Nawegam NP.
14.	Dry Sal bearing Forest	–	Amarkamlak–Achanakamar WLS in MP.
15.	Northern Dry Mixed Deciduous Forest	–	Several.
16.	Anogeisser Pendula Dry Deciduous Forest	–	Kumbhalgarh Sariska, Sibpur, Ranthaublue
17.	Hardwickia Dry Deciduous Forest	–	Nagarjun Sagar, Sh. Benkatsaran.
18.	Southern Tropical Thorn Forest	–	Chincholi WLS.

	Forest Types		National Park/NLS
19.	Northern Tropical Thorn Forest	–	Gir NP.
20.	Central Indian Sub-Tropical Broadleaf Hill Forest.	–	Ananlagiri WLS.
21.	East Sub-Tropical Broadleaf Wet Hill Forest	–	Mawamai, Norkrek, in Meghalaya.
22.	Himalayan Sub-Tropical Fine Forest	–	Ranikhet, Sudhatoli WLS.
23.	Southern Montane Wet Temperate Forest	–	Karimpuzha, Anamudi, Agastyamelai NPS.
24.	Eastern Montane Wet Temperate Forest	–	Namdhapa N.P.
25.	Western Himalayan Moist Temperate Forest	–	Kedarnath NP.
26.	West Himalayan Sub-Alpine Forest/Pasture	–	Dachigam
27.	East Himalayan Sub-alpine Forest/Pasture	–	Kanchandzonga, Nandhapa.
28.	Himalayan Moist Alpine Forest/Pasture.	–	Daclingaon, Kanchandzonga, Kedarnath, G.P. Vihar.
29.	Desert Grassland & Scrub	–	Desert W.Ls.
30.	Alluvial grassland	–	Kaziranga, Manas, D, Ering, NPS.
31.	Rivers, Riverine Habitat	–	National Chambal, Nagarjun Sagar, Satkosir NPS.

(*iii*) **Threatened mammals in various Protected areas**

	Species	Significant Existing Protected area	Major Areas now Proposed
1.	Andaman pig	Detailed distribution not known.	-
2.	Bharal	Hemis, Nanda Devi, Kanchandzonga.	Ladakh, Spiti, Pindan, Sikkim.
3.	Binturong	Kanchandzonga, Moiling, Namdapa.	Pangola in Sikkim, several in Arunachal.
4.	Black buck	Velavador, Point Calimere, Tall.	Upgrading closed areas of Rajashtan.
5.	Brow-antlered deer	Keibul-Jamgar NP.	-
6.	Capped langur	Kaziranga NP.	New park in Assam, Arunachal.
7.	Coracal	Sariska, Panna N.P.	Several new areas.
8.	Chinkara	Desert N.P., Panna N.P.	New areas in Maharashtra.
9.	Crab-eating macaque	None	Great Nicobar.
10.	Desert cat	Low density in desert	New N.Ps. in Gujarat.
11.	Dugong	Marine N.P. at Gujarat, Andaman & Nicobar islands.	Gulf of Manner & in TN.
12.	Fishing Cat	Kaziranga N.P.	-
13.	Chowsarga	Gir N.P.	-
14.	Golden langur	Manas W.L.S. to Assam.	-
15.	Ibex	–Hemis, Gt. Himalayan, Smaller in H.P.	-
16.	Himalayan Tahr	Gt. Himalayan N.P.S	Pindai W.L.S. in U.P.
17.	Hispid hare	Mana W.L.S.	Upgrading Mana.
18.	Hoolock gibbin	Nandhipa, Balphakran.	Several in Arunachal, Dzuko, Dkansiri.
19.	Indian wolf	Hemis, Kanchendzonga.	-
20.	Hargal	Dacligaon.	-
21.	Phayre's leaf monkey	Tripura	Trishna W.L.S.
22.	Red panda	N.E. & W.L.S.	New W.L.S.

	Species	Significant Existing Protected area	Major Areas now Proposed
23.	Lion tailed macaguse	Anamalar, Mindan thurai	
24.	Lynx	Eranikulam.	
25.	Malabar civet	Recent collection Kerala.	-
26.	Malay sunbear	Possibly in Balphekram NP.	-
27.	Markhor	Not found in any NL & WLS	-
28.	Mouse deer	Widespread in Deccan.	
29.	Musk deer	Great Himalayan NP, Kedarnath WLS	- Upgraded areas.
30.	Nilgiri thar	Sikkim, Arunachal.	- Upgraded.
31.	Palla's cat	Absence of reliable information.	
32.	Pigmi hog	Manas W.L.S.	- Upgraded.
33.	Rhinoceros.	Kaziranga ND, Jaldapara WLS.	-
34.	Rusty spotted cat	Vanada WLS.	-
35.	Serow		-
36.	Sloth bear	Kanha NP, Ranthambhore NP.	-
37.	Snow leopard	Hemis, Great Himalayan NP.	- Ladakh WLS, Hemis & Siachar-Shyok NP.
38.	Swamp deer	Dadica, Kaziranga, Kanha NPS.	-
39.	Takir	Arunachal north of Lohat	- Dibang valley walong.
40.	Tibetan gazelle	-	
41.	Tiger	- Tiger Reserve	-
42.	Urial.	- Hemis N.P.	-
43.	Wild buffalo.	- D. Ering WLS, Kaziranga NP, Indravati WLS.	-
44.	Gaur.	- Kanha ND, palaman NP.	- Melghat WLS.

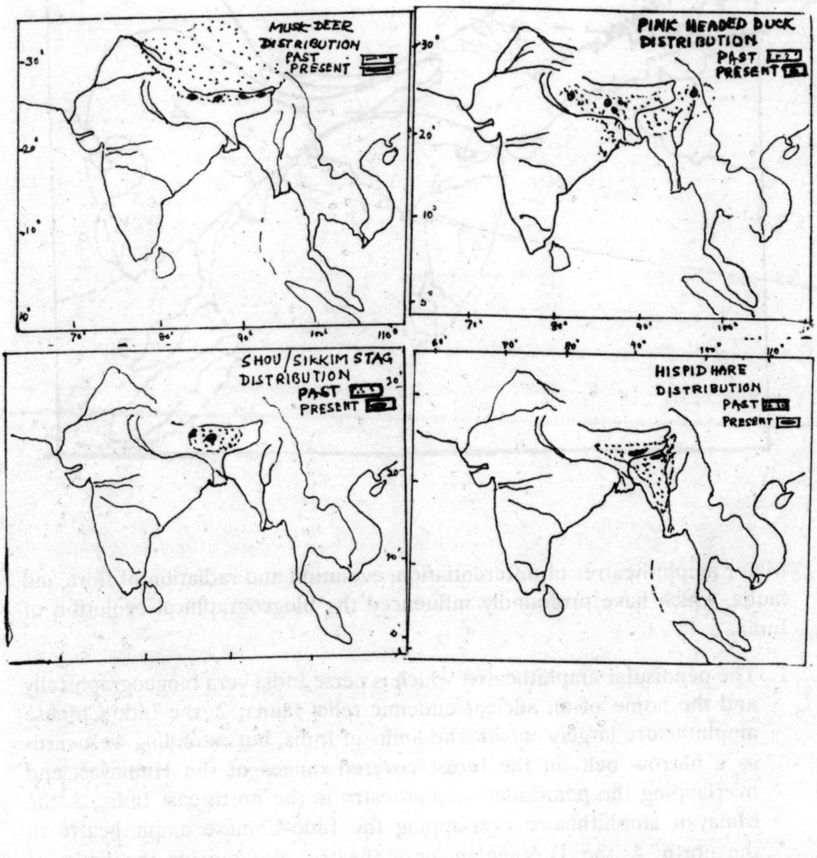
MUSK DEER
DISTRIBUTION
PAST
PRESENT
PINK HEADED DUCK
DISTRIBUTION
PAST
PRESENT
SHOU/SIKKIM STAG
DISTRIBUTION
PAST
PRESENT
HISPID HARE
DISTRIBUTION
PAST
PRESENT

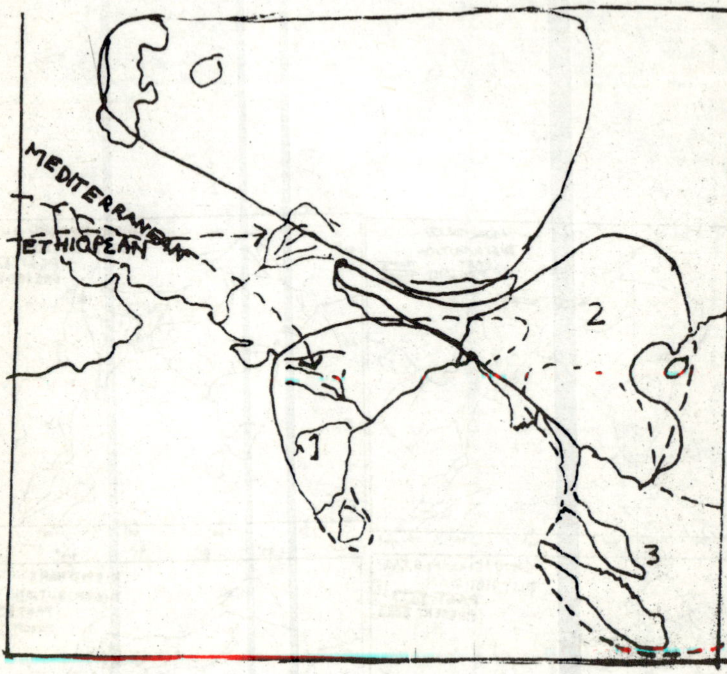

Major amphitheatres of differentiation, evolution and radiation of flora and fauna, which have profoundly influenced the biogeographical evolution of India.

1. The pennisular amphitheatre, which is perse India vera biogeographically and the home of an ancient endemic relict fauna; 2. the Indo-Chinese amphitheatre largely outside the limits of India, but extending westwards as a narrow belt on the forest covered ranges of the Himlayas, and overlapping the pennisular amphitheatre in the north-east India; 3. the Malayan amphitheatre overlapping the Indo-Chinese amphitheatre in the north; 4. the Turknenian amphitheatre, also outside the limits of India, except for a narrow southern fringe encroaching on the higher Himalayas above the forest line and an area of differentiation of young steppes fauna. The amphitheatre 2 and 3 are regions of differentiation of young tropical forest faumas. The amphitheatre, 2-4 represent the biogeographical appendages of India.

(*iv*) **Principal Wildlife Species in Wildlife Sanctuaries and National Parks**

Sl.No.	Sanctuary/ National Park	Location	Principal Animal/Birds (Fauna) Available
Andhra Pradesh			
1.	Pakhal Wildlife Sanctuary (878 sq.km)	Warangal Distt 60 km from Warangal	Tiger, Panther, Sambar, Chital, Nilgai, Hyaena, Wild spotted deer, duck, storks.
2.	Eturnaharam Wildlife Sanctuary (803 sq. km.)	Warangal Distt 80 km from Kazipet	Tiger, panther, gaur, jungle cat, sambar, black-buck, walbes, jackals.
3.	Pocharam Wildlife Sanctuary (130 sq. km.)	Medak District 1km from Medak.	Panther, chital, chinkara, peafowl, and water birds, wildbear, flamingoes.
4.	Kawal Wildlife Sanctuary (893 sq. km.)	Adilabad District 50 km from Mancherial Railway Stn.	Tiger, panther, gaur, sambar, chital, black buck, wild barking deer etc.
5.	Kolleru Pelicanary (673 sq.km).	West Godavari.	Pelicans, glamingo, heron, painted stork, avocet, teal, and tern, Gargang Teals, Pochards spot hills, grey peli cray etc.
6.	Neelapattu Bird	Nellore-Distt	Grey pelican, heron, cormorant, teal, duck etc.
7.	Kinnersani Wildlife Sanctuary (878 sq. km.)	Khammam-Distt Paloncha Division	Tiger, panther, wolf, gaur, chital, sambar, nilgai, sloth bear.
8.	Lanjamdugu (Sivaram) Sanctuary (30 sq. km.)	50 km from Mancherial in Adilabad-Distt.	Tiger, panther, chital, sambar, nilgai, langur, sloth bear and wild bear, pythons monkeys, crocodiles, etc.
9.	Manjira Sanctuary (20 sq. km.)	20 km from Singereddi Dam to Manjira Barrange-near Sangareddi.	Marsh Crocodiles, Water Birds

Sl.No.	Sanctuary/ National Park	Location	Principal Animal/Birds (Fauna) Available
10.	Papikonda Sanctuary (591 sq. km.)	Trijunction of Kamma, East and West Godavari border along river Godavari.	Tiger, panther, wild dog, hyaena, jackal, wolves, gaur, four horned antelope, chital, sambar and nilgai.
11.	Corinaga W.L. Sanctuary (236 sq. km.)	Kakinada.	Estuarine crocodile, Otter fishing cat, jackal, sea-gulf pelican, stork, herons, flamingoes, ducks, etc.
12.	Nagarjun Sagar Srisailam Sanctuary (1347 sq. km.)	100 km. south to Hyderabad.	Tiger, panther, sloth bear, Wild bear, chital, sambar, nilgai, black buck, jackal, fox, wolf and mugger crocodile.
13.	Pulicat Sanctuary (600 sq.km.)	Nellore-Distt	Flamingoe, pelican, duck, teal, storks, crane, bear, heron.
14.	Pranahita Wildlife Sanctuary	Adilabad Distt.	Black buck, tiger.
15.	Mahavira Herina Vanasthali.	33 km to Hyderabad Vijaywada National Highway	Gargang, teals,pochards, spot bill, grey pelican.

Andaman & Nicobar

N.B. Uptodate lists of National park and Sanctuaries may be seen in the Text under Conservation of Andaman & Nicobar Islands.

Arunachal Pradesh (83,000 sq.km)

1.	Namdapha National Park (1807 sq. km.)	District tirap	Tiger, panther, snow leopard, clouded leopard, golden cat, binturang, wild, dog, wild buffalo, gaur, hog deer, barking deer, elephant, sloth bear, paython, king cobra, Indian cobra, montior lizard.

Sl.No.	Sanctuary/ National Park	Location	Principal Animal/Birds (Fauna) Available
2.	Lali Wildlife Sanctuary.	District Siang	Wild buffalo, tiger, panther, hog deer, elephant, python, barking deer, sambar, water birds.
3.	Itanagar Wildlife Sanctuary (140 sq. km.)	District Subansiri.	Elephant, gaur, tiger, panther, sambar, hog deer, barking deer, sloth bear
4.	Pakkui Wildlife Sanctuary (862 sq. km.)	District Kaneng	Elephant gaur, sambar, hog deer, barking deer, python.

Others–Moiling National park (183 sq.km), Mahao Wildlife Sanctuary (282 sq.km), D' Ering (190 sq.km) Wildlife Sanctuary. These have Sikkim Stag, wild buffalo, rhinos, etc.

Assam

Sl.No.	Sanctuary/ National Park	Location	Principal Animal/Birds (Fauna) Available
1.	Garampani Wildlife Sanctuary	District Diphu	Elephant, wild buffalo, Tiger, leopard, common langur, hoolock
2.	Kaziranga National Park (696 sq. km.)	Jorhat	Great Indian one-horned rhinoceros, Wild buffalo, elephant, gaur, leopard, cat, wild bear, civet cat, otter, swamp deer, hog deer, sambar, tiger, python, pelican, partridge, florican
3.	Lakhawa Wildlife Sanctuary (70 sq. km.)	Nowgong	Rhinoceros, wild buffalo, swamp deer, hog deer, water ducks, cormorants.
4.	Manas Wildlife Sanctuary (390 sq. km.)	Barpeta	Elephant, tiger, panther, gaur, wild buffalo, great Indian one-horned rhinoceros, golden langur, civet cat, otter, swamp deer, hog deer, sambar, pigmy hog, water monitor, wild boar, great pied hornbill, florican, Merganser, Wild dog.

Sl.No.	Sanctuary/ National Park	Location	Principal Animal/Birds (Fauna) Available
5.	Orang Wildlife Sanctuary (92 sq. km.)	Tezpur	The great Indian one-horned rhinoceros, elephant, leopard, sambar, barking deer, swamp deer, tiger, water birds, green pigeon, florican, quail, whistling teal, geese, hawks.
6.	Poha Wildlife Sanctuary (92 sq. km.)	North Lakhimpur.	Wild buffalo, barking deer.
7.	Sonai-Rupa Sanctuary	Tezpur	The great Indian one-horned rhinoceros, elephant, wild buffalo, sloth bear, wild dog, wild boar, sambar, barking deer, swamp deer leopard, tiger horn-bill, imperial pigeon etc.
8.	Pabitora		Rhinoceros, wild buffalo, hog deer, water ducks, cormorants.
Bihar			
1.	Hazaribagh Wildlife Sanctuary (86 sq. km.)	Hazaribagh	Tiger, leopard, sambar, chital, nilgai, wild boar, wild cat, peafowl etc.
2.	Bhmbandh Wildlife Sanctuary (682 sq. km.)	Monghyrul	Tiger, leopard, sambar, chital, wild boar, wolf, water birds.
3.	Mahuadana Wildlife Santuary (51 sq. km.)	Deltongunj	Tiger, leopard, wolf, chital, barking deer, wild boar etc.
4.	Dalma Wildlife Sanctuary (193 sq. km.)	Singhbhum.	Elephant, leopard, wild boar, mouse-deer, barking deer, barking sloth bear.

Sl.No.	Sanctuary/ National Park	Location	Principal Animal/Birds (Fauna) Available
5.	Palamau National Park (213 sq. km.)	Daltongunj.	Elephant, panther, leopard wild boar, barking deer, gaur, chital, sambar, peafowl etc.
6.	Gautam Budha Wildlife Santuary (260 sq. km.)	Gaya	Tiger, leopard, sambar, chital, barking deer, peafowl etc.
7.	Rajgir Wildlife Sanctuary (36 sq. km.)	Patna.	Tiger, leopard, sambar, chital, barking deer, peafowl etc.
8.	Topechanahi Wildlife Sanctuary (12 sq. km.)	Dhanbad.	Variety of ducks.
9.	Lawalang Wildlife Sanctuary (207 sq. km.)	Hazaribagh	Tiger, leopard, sloth bear, sambar, chital etc.
10.	Lanaul Madanpur & Udaipur Wildlife Sanctuary (9 sq. km.)	Champaran	Tiger, leopard, sambar, chital, wild boar, nilgai, wolf etc.
11.	Kaimur Wildlife Sanctuary (1342 sq. km.)	Rohtas	Tiger, leopard, chinkara, sambar nilgai and crocodiles.
12.	Bamiaburu Wildlife Sanctuary	Singhbhum	Tiger, panther, sambar, elephant wild boar etc.
13.	Tabo Wildlife Sanctuary	Singhbhum	-do-
14.	Kodaram Wildlife Sanctuary (207 sq. km.)	Hazaribagh.	Tiger, leopard, sloth bear, sambar, chital four-horned, antelope.
15.	Valmiki Nagar Wildlife Sanctuary (462 sq. km.)	-	-
Goa, Daman and Diu			
1.	Mollam Wildlife Sanctuary (240 sq. km.)	Goa	Gaur, Sambar, mouse deer, barking deer, panther, flying squirrel, porcupine, civet, cat, ant-eater, slender loris, grey jungle fowl.

Sl.No.	Sanctuary/ National Park	Location	Principal Animal/Birds (Fauna) Available
2.	Bondla Wildlife Sanctuary (8 sq.km.)	Bondla Goa	-do-
3.	Cotigoa Wildlife Sanctuary (115 sq.km).	Goa	-do-
Gujarat			
1.	Gir National Park (259 sq.km.)	Junagarh	Asiatic lion, panther, striped hyaena, sambar, nilgai, chital, four-horned antelope, chinkara, wild boar, crocodile.
2.	Velavadar National Park (35 sq.km.)	Bhavnagar	Black buck, wolf.
3.	Nal sarovar Bird Sanctuary (765 sq.km)	Ahmedabad	Water birds.
4.	Wild Ass Sanctuary in and around the little Runn of Kutch. (4953 sq. km.)	Surendranagar	Wild ass, nilgai, wolf, chinkara.
Haryana			
1.	Sultanpur Lake birds Sanctuary (1 sq. km.)	Gurgaon.	Sarus carne, spot-bill, ruddy shell duck etc.
Jammu & Kashmir			
1.	Dachigam National Park (141 sq. km.)	Srinagar	Leopard, black bear, brown bear, baboon, serow, musk deer, hangual.
2.	Raj Parian Wild-life Sanctuary	Anantnag	Brown bear, black bear, musk deer, scrow, hangual, and chakor etc.

Sl.No.	Sanctuary/ National Park	Location	Principal Animal/Birds (Fauna) Available
3.	Shankarachari Wildlife Sanctuary.	Srinagar	Chakor
4.	Chummai Basin Wildlife Sanctuary	Anantnag	Brown bear, black bear, leopard, musk deer, etc.
5.	Hemis National Park (600 sq. km.)	Ladakh	Snow leopard, Pallas cat, Brown bear, Indian wolf, Tibetan ass, wild yak, Tibetan gazell, Tibetan antilope, urial, argali, marcopolo sheep, markhor, ibex, blue sheep, black naked crane
6.	Kistwar National Park (425 sq. km.)	Ladakh	
Himachal Pradesh			
1.	Great Himalayan Saichu Tuan National Park (1716 sq. km.)	Chamba.	Ibex, tahr, musk deer, brown bear, black bear, goral, serow, monal, trahopan, snow cock, kalij, Koklass and chakor, snow leopard, lynx, marten fox, panther, marmots, weasels.
2.	Gamgul-Siya Behli Wildlife Sanctuary (9 sq. km.)	Chamba	Tahr, brown bear, musk deer, goral, Kashmir stag, koklass, chakor, black bear, monal, tragopan, snow, cock, kalij, panther, weasel, marten, civet, fox, flying fox.
3.	Tundah Wildlife Sanctuary (64 sq. km.)	Chamba	Tahr, brown bear, black bear, musk deer, goral, serow, ibex, monal, tragopan, snow cock, kalij, chakor, and chir, snow leopard, panther, snow panther, weasel, martens, civet, fox, flying fox

Sl.No.	Sanctuary/ National Park	Location	Principal Animal/Birds (Fauna) Available
4.	Kugti Wildlife Sanctuary (118 sq. km.)	Chamba	Tahr, brown bear, black bear, musk deer, goral, serow, ibex, monal, tragopan, snow cock, kalij, koklash chakor, panther, snow leopard, marten, civet, fox, flying fox.
5.	Kelatop and Khajjar Wildlife Sanctuary (47 sq. km.)	-do-	Goral barking deer, serow, black bear, koklash, kalij, chir, panther, snow leopard, marten, civet, fox, flying fox.
6.	Nergu and Winch Camo W.L.S. (278 sq. km.)	Mandi	Goral, black bear, serow, musk deer, monal, kalij, koklash and chir, panther, snow leopard, marten, civet, fox flying fox.
7.	Shikari Devi WLS. (214 sq. km).	-do-	Black bear, goral, barking deer, monal, Kalij, chir, chakor, partridge, panther, show leopard, marten, civet, fox, flying fox.
8.	Bandli Wildlife Sanctuary (41 sq. km.)	Mandi	Goral, barking deer, black bear, kalij, chakor, partridge panther, snow leopard, marten, civet, fox, flying fox.
9.	Gobindsagar Wildlife Sanctuary (100 sq. km.)	Bilaspur	Duck, teal, goose, crane.
10.	Shri Naina W.L.S. (156 sq.km.)	-do-	Sambar, barking deer, black bear, wild boar, nilgai, kalij, red jungle fowl, partridge, panther, snow leopard, marten, civet, fox, flying fox.

Sl.No.	Sanctuary/ National Park	Location	Principal Animal/Birds (Fauna) Available
11.	Majathal and Harseng Wildlife Sanctuary. (92 sq.km).	Solan	Goral, barking deer, black bear, kalij, red jungle fowl, partridge, panther, snow leopard, marten, civet, fox, flying fox.
12.	Baranghati I & II WLS (42 sq.km).	Kinnaur	Goral, black bear, serow, monal, kalij, koklash, chakor, panther, civet marten, flying fox.
13.	Shilli W.L.S. (2 sq.km).	Solan	Barking deer, chakor, kalij, partridge, panther, snow leopard, marten, civet, fox, flying fox.
14.	Water supply catchment area Simla (10 sq.km).	Simla	Goral, barking deer, serow, kalij, koklash, monal (rare), panther, snow leopard, marten, civet, fox, flying fox.
15.	Talra Wildlife Sanctuary (26 sq.km)	Simla	Goral, black bear, sambar, musk deer, Kalij, koklash, chakor, monal, panther, flying fox, civet, marten
16.	Darlaghat W.L.S. (140 sq.km).	Solan	Goral, barking deer, chakor, kalij, partidge, panther, snow leopard, marten, civet, fox, flying fox.
17.	Raksham Chitkul W.L.S. (3.8 sq.km).	Kinnaur.	Bharal, goral, black bear, brown bear, monal, koklash, snow, cock, chakor, panther, fox, marten, flying fox.
18.	Lipa Asrang W.L.S. (109 sq.km.)	Kinnaur.	Ibex, baral goral, monal, koklash, snow cock, chakor, panther, fox, marten, flying fox.
19.	Renuka W.L.S. (4 sq.km.)	Sirmur.	Sambar, goral, barking deer, wild boar, kalij, peafowl, red jungle fowl, partridge, panther, fox, marten, civet, flying fox.

Sl.No.	Sanctuary/ National Park	Location	Principal Animal/Birds (Fauna) Available
20.	Simbalbara W.L.S. (19 sq.km.)	Sirmur	Sambar, goral, barking deer, wild boar, kalij, peafowl, red jungle fowl partridge, panther, fox, marten, civet, flying fox, chital and monal.
21.	Manali W.L.S.	Kulu	Tahr, serow, goral, black bear, kalij, koklash, chakor, partridge, chir, panther, civet, marton, flying fox.
22.	Kias W.L.S.	-do-	-do-
23.	Khokhana W.L.S. (23 sq.km.)	-do-	-do-
24.	Chail W.L.S. (23 sq.km.)	Solan.	Goral and barking deer.
25.	Thirhan W.L.S.	Kulu.	
Karanataka			
1.	Bandipur National Park (874 sq.km.)	Mysore	Indian elephant, tiger, gaur, sambar, chital, barking deer, wild dog, wild boar, jackal, sloth bear, panther, four-horned antelope, malabar squirrel, jungle fowl, partridge, green pigeon, bush quail, etc.
2.	Bannarghatta National Park (104 sq.km.)	Bangalore	Elephant, sloth bear, chital, barking deer, grey partridge, bush quail, jungle fowl etc.
3.	Nagerhole National Park (572 sq.km).	Coorg	Elephant, tiger, panther, chital, sambar, sloth bear, jungle fowl, partridge etc.
4.	Someshwara Wildlife Sanctuary (88 sq.km).	South Kanara	Chital, gaur, sambar, panther, tiger, sloth bear, wild dog, etc.

Sl.No.	Sanctuary/ National Park	Location	Principal Animal/Birds (Fauna) Available
5.	Brahmagiri Wildlife Sanctuary	Coorg	Tiger, panther, sambar, chital, barking deer, sloth bear, cobra, flying squirrel, malabar squirrel, civet cat, mouse, deer.
6.	Malkote Temple Wild-life Sanctuary (50 sq.km).	Mandya	Panther, wolf, sloth bear, black-buck, hare, porcupine, wild boar.
7.	Ghataprabha Bird Sanctuary (30 sq.km.)	Belgaum	Egret, cormorant, heron etc.
8.	Nugu Wildlife Sanctuary (30 sq.km.)	Mysore	Elephant, tiger, chital, sambar, barking deer, gaur, wild boar, sloth bear etc.
9.	Tungabhadra Wildlife Sanctuary.	Bellary	Black-buck, chinkara, chital, panther,pigs, four-horned antelope fox, grous, florican, great Indian bustard, sloth bear, sambar etc.
10.	Ranebennur Black Buck Sanctuary	Dharwar	Black buck, chital, wolf, hare, great Indian bustard.
11.	Mukambika Wildlife Sanctuary (100 sq.km.)	South Kanara	Panther, sambar, elephant, wild boar, porcupine.
12.	Sharawathy Valley	Shimoga	Indian elephant, tiger, panther, gaur, sambar, chital, barking deer, wild dog, wild boar, jackal, sloth beer, four-horned antelope, malabar squirrel, jungle fowl, partridge, green pigeon, bush quail, owl, lion, tailed monkey, tortoise,crcodile, python, cobra, krait, viper, green snake etc.

Sl.No.	Sanctuary/ National Park	Location	Principal Animal/Birds (Fauna) Available
13.	Biligiri Rangaswamy Wildlife Sanctuary.	Mysore	Elephant, gaur, chital, barking deer, sambar, panther, sloth bear.
14.	Bhadra Wildlife Sanctuary (492 sq.km)	Chikmangalore	Gaur, elephant, panther, wild boar, sloth bear, sambar, chital, barking deer.
15.	Shattihalli Wildlife Sanctuary (396 sq.km)	Shimoga	Gaur, elephant, panther, wild boar, sambar, chital, barking deer.
16.	Ranganthitoo Bird Sanctuary (1 sq.km).	Mysore	Openbill stork, white ibis, little egret, cattle egret, dartar, cormore ant, pond heron, river tern, spoon bill, crocodile etc.
17.	Dandeli Wildlife Sanctuary (5730 sq.km)	Dharwar	Tiger, panther, elephant, gaur, sloth bear, sambar, chital, wild boar etc.
Kerala			
1.	Eravikulam Rajmallay National Park.	Iddukki	Elephant, gaur, sambar, barking deer, tiger, panther, civet, jungle cat, nilgai, langur, wild boar, lion- tailed macaque, malabar squirrel,wild dogs, nilgiri, tahr, tiger, imperial pigeon, grey jungle fowl.
2.	Periyar Wildlife Sanctuary (427 sq.km)	Iddukki	Elephant, tiger, panther, wild dog, gaur, sloth bear, nilgai, wild boar, sambar and barking deer.
3.	Parambikulam Wildlife Sanctuary (285 sq.km.)	Palghat	Elephant, gaur, leopard, tiger, sloth bear, nilgai, chital, sambar, crocodile.

Sl.No.	Sanctuary/ National Park	Location	Principal Animal/Birds (Fauna) Available
4.	Neyyar Wildlife Sanctuary (128 sq.km.)	Trivandrum.	Elephant, tiger, leopard, wild bear, sambar, chital, nilgai, tahr.
5.	Prechi-Vazahani Wildlife Sanctury (125 sq.km.)	Trichur.	Elephant, wild boar, chital.
6.	Wynard Wildlife Sanctuary (344 sq.km)	Cannanore & Kozhikode.	Elephant, gaur, sambar, wild boar, chital, barking deer etc.
7.	Idukki Wildlife Sanctuary (70 sq.km)	Iddukki	Elephant, sambar, sloth bear, wild bear, wild boar, gaur etc.

N.B. Others are Aralem (55 sq.km), Chimnomy (90 sq.km.) Chinner (90 sq.km Peppara (53 sq.km).

Madhya Pradesh

Sl.No.	Sanctuary/ National Park	Location	Principal Animal/Birds (Fauna) Available
1.	Kanha National Park (940 sq.km).	Mandla & Balaghat.	Tiger, panther, gaur, barasingha, chital, sambar, black buck, chow-singha, barking deer, mouse, deer, nilgai, wild dog, boar.
2.	Bandhagarh National Park (449 sq.km).	Shahdol	Tiger, panther, gaur, chital, sambar, nilgai, chinkara, barking deer, bear, wild boar and a variety of upland birds.
3.	Shivpuri National Park.	Shivpuri.	Tiger, panther, sloth bear, hyaena, sambar, spotted deer, four-horne antelope, black buck, nilgai, chinkara, wild boar, crocodile etc.
4.	Bori Wildlife Sanctuary (483 sq.km.)	Hoshangabad	Tiger, panther, bison, sambar, chital, nilgai, barking deer, chinkara, wild boar, bear.

Sl.No.	Sanctuary/ National Park	Location	Principal Animal/Birds (Fauna) Available
5.	Kutree Wild Buffaloes Game Sanctuary.	Baster	Wild buffalo, tiger, panther, sloth bear, nilgai, chital, sambar, wild pigs, barking deer, bison, wild dogs, chowsingha, hill myna etc.
6.	Tamor-pigla WLS (609 sq.km.)	Sarguja	Tiger, panther, gaur, cheetal, sambar etc.
7.	Semarsot Wildlife Sanctuary (430 sq.km.)	-do-	-do-
8.	Panna National Park (543 sq.km.)	Panna	Tiger, panther, sambar, chital, wild pig, nilgai, peafowl.
9.	Gomardha Wildlife Sanctuary (278 sq.km.)	Raigarh	Tiger, panther, sambar, chital, nilgai, etc.
10.	Badalkhol Wildlife Sanctuary (104 sq.km)	Raigarh	Tiger, panther, sambar, chital, nilgai etc.
11.	Barnawapara Wildlife Sanctuary (247 sq. km)	Raipur	Tiger, panther, gaur, sambar, chital, bear etc.
12.	Sirper Bird Sanctuary	Raipur	Variety of upland birds and multitude of birds along with wild animals
13.	Sitanadi Wild-life Sanctuary	Raipur	Tiger, panther, sambar, chital, bison, wild boar, barking deer, pea-cock etc.
14.	Noradehi Wildlife Sanctuary (1187 sq. km)	Sagar Damoh & Narsingh-garh	Tiger, panther, sambar, chinkara, chital, nilgai, bear etc.

Sl.No.	Sanctuary/ National Park	Location	Principal Animal/Birds (Fauna) Available
15.	Narsinghgarh Game Sanctuary (57 sq. km)	Raigarh	Panther, sambar, chital, wild pig, peacock, nilgai
16.	Sanjay – Dubari Wild-life Sanctuary (365 sq. km)	Sidhi	Tiger, panther, sambar, chital, nilgai, barking deer, chinkara etc.
17.	Bagdara Wildlife Sanctuary (479 sq. km)	Sidhi	Panther, black. buck, chinkara, sambar, nilgai, wild pig etc.
18.	Panchmarhi Wildlife Sanctuary (461 sq km)	Hoshangabad	Tiger, panther, bear, bison, spotted deer, sambar, barking deer, nilgai, etc.
19.	Achanakmar Game Sanctuary	Bilaspur	Tiger, bison, sambar, chital, boar, peacock etc.
20.	Ratapani Wildlife Sanctuary (689 sq km)	Raisen	Tiger, panther, sambar, chital, blue bull, chinkara etc.
21.	Pench National Park (293 sq km)	Chnidwara & Seoni	Tiger, panther, gaur, chital, sambar, nilgai, etc.
22.	Singhari Wildlife Sanctuary (288 sq km)	Raisen	Tiger, panther, gaur, chital sambar, nilgai, etc.
23.	Gandhi Sagar	Mandsaur	Water birds, chital, sambar, chinkara, barking deer.
24.	Kheoni Wildlife Sanctuary (123 sq km)	Dewas	Panther, chital, sambar, nilgai, bear etc.
25.	National Chambal Wild-life Sanctuary (320 sq km)	Morena	Charial

N.B. : Others : Ken Ghareil (45 sq.km), Son Ghareil (105 sq. km).

Sl.No.	Sanctuary/ National Park	Location	Principal Animal/Birds (Fauna) Available
MAHARASHTRA			
1.	Tadoba National Park (117 sq km)	Chandrapur	Tiger, panther, sloth bear, gaur, sambar, chital, nilgai, chinkara and crocodile
2.	Pench National Park (267 sq.km.)	Nagpur	Tiger, panther, gaur, sloth beer, sambar, chital, barking deer, four-horned antelope, nilgai, chinkara, jungle fowl and peafowl
3.	Nawegaon National Park (134 sq. km.)	Bhandara	Tiger, panther sloth beer, gaur, sambar, chital, barking deer, nilgai and migratory birds.
4.	Boriveli National Park	Bombay Suburban	Panther, sambar, four-horned antelope, mouse deer, wild boar and langur
5.	Dhakna-kolkaz Wildlife Sanctuary	Amravati	Tiger, panther, gaur, sloth bear, sambar, barking deer, four- horned antelope, wild boar, chital, bird life is plentiful
6.	Yawal Wildlife Sanctuary (178 sq. km.)	Jalgoan	Panther, jungle cat, nilgai, bonnet macaque and common langur, wild dog, sambar, lyaena, tiger, chital and chinkara, peafowl, grey jungle fowl, green pigeon, sand grouse and partridge
7.	Radhanagari Wildlife San-tuary (72 sq. km.)	Kolhapur	Gaur, panther, sambar, and wild boar, bird life is fairly levish
8.	Karnala Wild Bird Sanctuary (4 sq. km.)	Raigad	Ashy minivet, paradise fly-catcher, shama, malabar whistling thrush, racket-tailed drongo, wood peaker etc. Other wildlife seen are the panther, four-horned antelope and common langur

Sl.No.	Sanctuary/ National Park	Location	Principal Animal/Birds (Fauna) Available
9.	Tansa Wildlife Sanctuary (305 sq. km.)	Thane	Panther, four-horned antelope, chital, sambar and wild boar, bird population is diverse and colourful around Tansa Lake
10.	Nagzira Wild-life Sanctuary (153 sq.km.)	Bhandara	Tiger, panther, bison, sloth bear, sambar, four-horned antelope, nilgai, chital, barking deer and mouse deer, birds too are colourful and varied the prominant being the peafowl and grey-jungle fowl
11.	Bor Wildlife Sanctuary (61 sq. km.)	Wardha	Panther, sloth bear, sambar, nilgai, chital, barking deer and wild boar, Tiger is occasionally seen
12.	Kinwat Wildlife Sanctuary	Yeotmal and Nanded	Tiger, panther, nilgai, sloth bear, sambar, chital, chinkara, barking deer and wild boar.
13.	Great Indian Bustard Wildlife Sanctuary (8496 sq. km.)	Ahmednagar	G.I.B.
14.	Deulgaon Dehekuri wild-life Sanctuary (2 sq. km.)		Black buck
MANIPUR			
1.	Keibul Lamjao National Park (40 sq. km.)	Central	Brow antelered, deer, wild goat, and water birds
MEGHALAYA			
1.	Balphakram National Park (220 sq. km.)	Garo hills	Wild elephant, tiger, buffalo, gibbon

Sl.No.	Sanctuary/ National Park	Location	Principal Animal/Birds (Fauna) Available
2.	Norkrek N.P. (50 sq km)	Garo hills	Wild elephant (Wild Citrus)
3.	Siju Wildlife Sanctuary (5 sq. km.)	Garo hills	Wild elephant (pitcher plant)
4.	Bahmara Wildlife Sanctuary	Garo hills	Wild elephant (Pitcher plant)
5.	Nong khyllem Wildlife Sanctuary (29 sq km)	Garo hills	Wild elephant, deer
MIZORAM			
1.	Dampa Wildlife Sanctuary	Aizawl	Elephant, tiger, leopard, sambar, barking deer, himalayan bear, wild boar, wild-dogs, wild-cat, gaur, leopard cat, king cobra, phython, hornbill pheasant etc.
NAGALAND			
1.	Intangki Wildlife Sanctuary (56 sq.km.)	Kohima	Gaur, boar, elephant, barking deer, wild bear, clouded leopard, panther tiger, pingolin and various kinds of birds and reptiles.
2.	Pulie Badze Wildlife sanctuary (9 sq km)	–do–	–do–
3.	Fakim Wildlife Sanctuary (6 sq. km.)	–do–	–do–
ORISSA			
1.	Satkosia Gorge Wildlife Sanctuary (478 sq. km.)	Dhenkanal Pruri, Cuttack and Phulbani	Gharial and mugger, tiger, leopard, jungle cat, civet, gaur, ratel, sloth bear, sambar, chital, nilgai four-horned antelope, hornbill, jungle fowl and other birds

Sl.No.	Sanctuary/ National Park	Location	Principal Animal/Birds (Fauna) Available
2.	Bhitar Kanika Sanctuary	Cuttack	Salt-water crocodile, leopard, hyena, jungle cat, leopard cat, wild boar, chital, sambar, giant, squirrel water monitor, sea-turtle, king cobra, python, painted stork, adjutant storks, open billed stork, white ibis, black ibis, etc.
3.	Chilka Sanctuary	Ganjam & Puri	Black buck, chital, ibis, crane, cormorant, agret, flamingo and pelican
4.	Kuldiha Wildlife Sanctuary	Balasore	Elephant, gaur, chital and sambar
5.	Karlapat Wildlife Sanctuary (255 sq km)	Kalahandi	Tiger, leopard, gaur, chital, sambar, etc.
6.	Saptasajya Wildlife Sanctuary	Dhenkanal	Elephant, gaur, leopard, sambar, chital
7.	Kapilash Wildlife Sanctuary	—do—	—do—
8.	Ushakothi Wildlife Sanctuary	Sambalpur	Elephant, gaur, tiger, leopard, sambar, chital and barking deer
9.	Similipal National Park (303 sq km)	Mayurbhanj	Tiger, elephant, gaur, chital, leopard, mouse deer, flying squirrel, and mugger
10.	Khalasuni Wildlife Sanctuary	Sambalpur	Elephant, gaur, chital, sambar, sloth bear
11.	Balukhand Wildlife Sanctuary	Puri	Black buck and chital
12.	Chandka Wildlife Sanctuary (230 sq. km.)	Puri	Elephant, panther, chital, sambar, barking deer, sloth bear, civet cat, etc.
13.	Debrigarh Wildlife Sanctuary (347 sq. km.)	Sambalpur	Leopard, sambar and gaur

Sl.No.	Sanctuary/ National Park	Location	Principal Animal/Birds (Fauna) Available
14.	Hadagarh	Keonjhar	–do–
15.	Nandankanan	Puri	It is a safari park
16.	Mahanadi & Baisipally	Nayagarh	Tiger, leopard, elephant, barking deer, sloth bear, civet cat
17.	Padmalota Wildlife Sanctuary	Phulbani	–do–
18.	Raihoda Wildlife Sanctuary	Angul	–do–
PUNJAB			
1.	Bir Moti Bagh Wildlife Sanctuary	Patiala	Nilgai, wild boar, hog deer, hare, peafowl, partridge, pigeon, dove, sparrow, babbler, wood pecker, myna and parakeet
2.	Bir Bhunnerheri Wildlife Sanctuary	Patiala	–do–
3.	Abohar Wildlife Sanctuary	Ferozepur	Black buck, hare, partridge, pigeon and dove
4.	Harike Headworks	Patiala	Duck, poacherd, teal, common coot, water hen, geese, nilgai and wild boar
RAJASTHAN			
1.	Ranthambhor National Park (392 sq. km.)	Sawai Madhopur	Tiger, panther, hyaena, jungle cat, civet, sambar, chital, nilgai, bear, wild boar, partridge, green pigeon, red spur fowl etc.
2.	Sariska Wildlife Sanctuary	Alwar	Tiger, panther, hyaena, jungle cat, civet, sambar, chinkara, nilgai, four-horned antelope, partridge, green pigeon, red spur fowl

Sl.No.	Sanctuary/ National Park	Location	Principal Animal/Birds (Fauna) Available
3.	Keoladeoghana National Park (28 sq. km.)	Bharatpur	Siberian crane, cormorant, storks spoon bill, quail, coot, heron, teal, tern, sambar, chital, black buck, wild boar, civet etc.
SIKKIM			
1.	Kanchandjunga National Park (849 sq. km.)	–	Ass, Wolf, bharal, argali, tahr, pallas cat, snow leopard, brown bear, serow goral, musk deer, red panda, cat, monal etc
UTTAR PRADESH			
1.	Corbett National Park (521 sq. km.)	Nainital/ Garhwal	Elephant, tiger, panther, sloth bear, nilgai, sambar, chital, wild boar, porcupine, peafowl, red jungle fowl, partridge and both the species of Indian inland crocodiles, goral and four-horned antelope
2.	Dudhwa National Park (490 sq. km.)	Lakhimpur Kheri	Tiger, panther, sloth bear, sambar, swamp deer, chital, hog deer, barking deer, nilgai, peafowl, jungle fowl and partridge etc.
3.	Govindpur Pashu Vihar (953 sq. km.)	Uttar Kashi	Brown and black himalayan bear, snow leopard, bharal, musk deer, tahr, serow, goral, panther, sambar, wild boar, monal pheasant, snow pigeon, green pigeon
4.	Nanda Devi National Park (630 sq km)	Chamoli	–do–

Sl.No.	Sanctuary/ National Park	Location	Principal Animal/Birds (Fauna) Available
5.	Rajaji Sanctuary	Saharanpur	Elephant, tiger, panther, sloth bear, nilgai, sambar, chetal, kakar, wild boar, porcupine, peafowl, red jungle fowl and partridge etc.
6.	Motichur Sanctuary	Dehradun	–do–
7.	Kishanpur Sanctuary	Lakhimpur Kheri	Brown and black himalayan bear, snow leopard, bharal musk deer, tahr, serow, goral, panther, sambar, wild boar, monal pheasant, snow pigeon, koklash, green pigeon etc.
8.	Kedarnath Wildlife Sanctuary (1067 sq km)	Chamoli	–do–
9.	Chandraprabha Sanctuary	Varanasi	Tiger, panther, sambar, Indian gazell, sloth bear, nilgai, peafowl, partridge sand grouse etc.
10.	Katarniaghat	Bahraich	Tiger, panther, sambar, bear, chital, black buck
11.	Ranipur Sanctuary	Banda	Tiger, panther, jungle cat, sambar, hyaena, fox, jackal, chital, chinkara, black buck etc.
12.	Chila Sanctuary	Garhwal	Tiger, panther, bear, elephant, chital, sambar, goral, nilgai, etc.
13.	Mahavir Swami Sanctuary	Lalitpur	Panther, nilgai, sambar, bear, chital, chinkara, black buck, black bear
14.	Chambal Sanctuary	-do-	Gharial, mugger

Sl.No.	Sanctuary/ National Park	Location	Principal Animal/Birds (Fauna) Available
15.	Valley of Flower National Park (88 sq.km.)	Garhwal	Magnificient assemblage of flowers; associated animals of the altitude, such as musk deer, Blue ship, tahr, snow partridge, snow cock etc.
WEST BENGAL			
1.	Lothian Island Wildlife Sanctuary	24-Parganas	Wild pigs, chital, otter, estuarine crocodile, ganhetic doplhin, water birds etc.
2.	Halliday Island	-do-	Royal bengal tiger, chital, water birds etc.
3.	Sajnakhali Wildlife Sanctuary	-do-	Tiger, wild boar, chital, cormonant, openbill, storks, snake, bird, white ibis, purple heron, grey heron, green bitter pelican, etc.
4.	Gorumara Jalpaiguri Wildlife Sanctuary	Jalpaiguri	Rhino, elephant, tiger, gaur, sambar, hog deer, wild boar and a variety of birds
5.	Chapramari Wildlife Sanctuary	-do-	Elephant, tiger, gaur, sambar, barking deer, wild boar and a variety of birds
6.	Jaldapara Wildlife Sanctuary	-do-	Rhino, elephant, tiger, leopard, wild boar, gaur, barking deer, hod deer, sambar, and a variety of birds
7.	Mahanadi Wildlife Sanctuary	Darjeeling	Tiger, elephant, gaur, sambar, hog deer, wild boar, gibbons and a variety of birds
8.	Sanchal Wildlife Sanctuary	-do-	Serow, Himalayan black baer, barking deer, goral and a variety of hill birds

Sl.No.	Sanctuary/ National Park	Location	Principal Animal/Birds (Fauna) Available
9.	Deer Parks Ballavpur	Birbhum	Black buck and chital
10.	Bethuadahari	Nadia	Chital, sambar and barking deer
11.	Parmandan	24-Parganas	Chital and mouse deer
12.	Sunderbans National Park (1330 sq. km.)	S. 24-Parganas	Tiger, different species of deer, wild boar, estuarine crocodile, gangetic dolphin
13.	Singalila N.P. (79 sq km)	Darjeeling	Himalayan bear, monal Kalij pheasant
14.	Neora N.P. (88 sq. km.)	Darjeeling	Leopard, wild boar, clouded leopard, panda, flying squirrel, barking deer, goral
CHANDIGARH			
1.	Sukhna Lake Wildlife Sanctuary	Chandigarh	Water birds, resident and migratory

(*v*) *Proposed Protected Areas*

(N.P. = National Park; W.L.S. = Wildlife Sanctuary; P.A. = Protected Areas).

ANDHRA PRADESH

(*a*) Nagarjun Sagar (N) N.P. (*b*) Nagarjun Sagar (S) N.P., (*c*) GBM WLS, (*d*) GBM N.P. (*e*) Mudinanipalli W.L.S., (*f*) Shri Venkateswara N.P., (*g*) Nachugunta-Kotta valem W.L.S., (*h*) Shri Harikota Island W.L.S., (*i*) Gudam-Marripakala W.L.S., (*j*) Gudam-Marripakala N.P., (*k*) Anantgiri W.L.S., (*l*) Donubhavi W.L.S., (*m*) Donubhavi N.P., (*n*) Navapada W.L.S., (*o*) Vellikonda W.L.S. and (*p*) Lankamalai W.L.S.

ARUNACHAL PRADESH

(*a*) Tawang W.L.S., (*b*) Tawang N.P., (*c*) Kalaktang W.L.S., (*d*) Lado W.L.S., (*e*) Lado N.P., (*f*) Pakhui N.P., (*g*) Palin W.L.S., (*h*) Tale

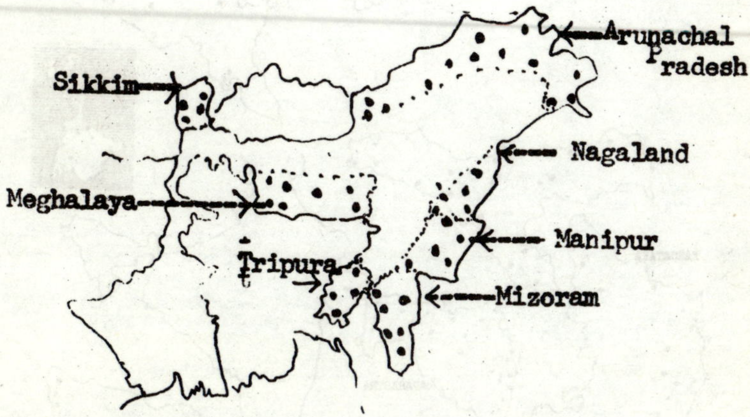

Fig. Northeastern states showing (in black dots) the National Parks, Wildlife sanctuaries and proposed protected areas

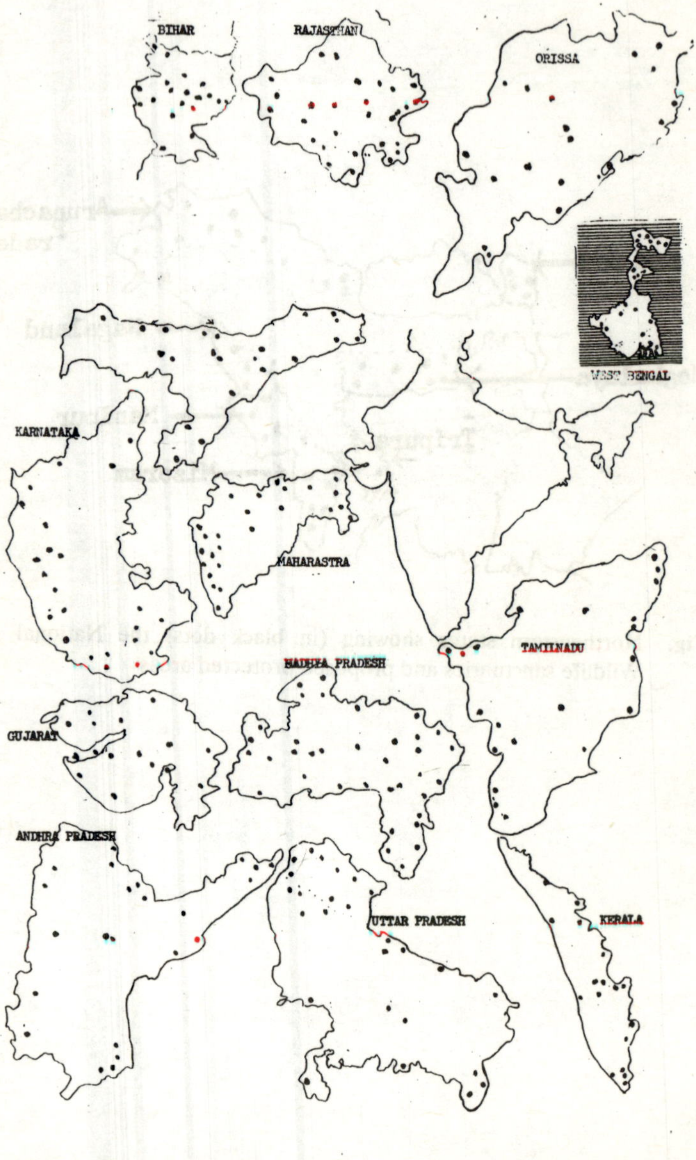

Fig. Principal states of India showing (black dots) the National Parks, Wildlife Sanctuaries and proposed protected areas to cover entire variety of wildlife species and varied forest types (Maps not to scale)

valley W.L.S., (*i*) Karsinganala W.L.S., (*j*) Moiling W.L.S., (*k*) D' Ening N.P., (*l*) Raneghat W.L.S., (*m*) Dibang valley W.L.S. (*n*) Dibang valley N.P. (*o*) Walong W.L.S., (*p*) Walong N.P., (*q*) Namdapha W.L.S. and (*r*) Tirap Evergreen N.P.

ASSAM

(*a*) Ripa-Chirang W.L.S., (*b*) Manas N.P., (*c*) Sonai-Rupai W.L.S., (*d*) Nameri W.L.S., (*e*) Karwa W.L.S., (*f*) Pobitra W.L.S., (*g*) Mikir hills W.L.S., (*h*) Hollongpahar W.L.S., (*i*) Desang mukh W.L.S., (*j*) Zamzing W.L.S., (*k*) Poba W.L.S., (*l*) Dibru W.L.S., (*m*) Digboi W.L.S., (*n*) Tinkhopani N.P., (*o*) Dhansirikak W.L.S., (*p*) Dhansirikak N.P., (*q*) Barail W.L.S., (*r*) Inner line Forest W.L.S. and (*s*) Barak W.L.S.

BIHAR

(*a*) Balmikinagar N.P., (*b*) Dalma N.P., (*c*) Saranda N.P. (*d*) Muta Mugger W.L.S., (*e*) Old Reserve W.L.S., (*f*) Ganga Dolphin W.L.S., (*g*) Kabartal N.P., (*h*) Kabartal W.L.S., (*i*) Kusheshwar W.L.S. and (*j*) Braital W.L.S.

GOA

(*a*) Mollem N.P.

GUJARAT

(*a*) Narayan Sarovar N.P., (*b*) Nal Sarovar N.P., (*c*) Great Rann N.P., (*d*) Banni W.L.S., (*e*) Mandvi W.L.S., (*f*) Wild Ass N.P., (*g*) Kori Creek W.L.S., (*h*) Cambay W.L.S. and (*i*) Purna W.L.S.

HARYANA

(*a*) Sultanpur N.P., (*b*) Damdama W.L.S., (*c*) Bhindawas W.L.S., (*d*) Nimbi W.L.S., (*e*) Semi Arid Area W.L.S., (*f*) Pinjore W.L.S. and (*g*) Kalesar W.L.S./N.P.

HIMACHAL PRADESH

(*a*) Spili N.P. and W.L.S. and (*b*) Saichu Tuan Nala N.P.

JAMMU AND KASHMIR

(*a*) Daultbag-Depsang N.P., (*b*) Shiang Shyok N.P./W.L.S., (*c*) Indus River W.L.S., (*d*) Leh-Kargil -- Srinagar Round W.L.S., (*e*) Suru W.L.S., (*f*) Rangdum W.L.S., (*g*) Num Kun W.L.S., (*h*) Hemis W.L.S., (*i*) Chang - Chenmo N.P., (*j*) Rupshee N.P./W.L.S. (*k*) Lahul-Kistwar W.L.S. (*l*) Saba-Chnkor W.L.S. and (*m*) Khandadhar W.L.S.

KARNATAKA

(*a*) Brahmagiri N.P., (*b*) Talakaveri W.L.S., (*c*) Pushpagin W.L.S., (*d*) Ammedikal W.L.S., (*e*) Kudremukh N.P., (*f*) Bhadra N.P., (*g*) Dandeli W.L.S., (*h*) Gangawalli-Angni W.L.S., (*i*) Chencholi W.L.S., (*j*) Kaveri W.L.S., (*k*) Nonavar W.L.S., (*l*) Pilarrkhan W.L.S., (*m*) Kundapur Mangrove W.L.S., (*n*) Rocky beach W.L.S. and (*o*) Bilgi W.L.S.

KERALA

(*a*) Karimpuzha N.P., (*b*) Shola Forest W.L.S., (*c*) Trivandrum Area W.L.S., (*d*) Palamala W.L.S., (*e*) Ponmuda (lower) W.L.S., (*f*) Ezhimala W.L.S., (*g*) Kurathimalai W.L.S., (*h*) Chirikala W.L.S., (*i*) Leathery Turtle Nesling beach W.L.S. and (*j*) Kumarkon W.L.S.

MADHYA PRADESH

(*a*) National Chambal N.P., (*b*) Choral W.L.S., (*c*) South Khargaon W.L.S., (*d*) Kalibhit N.P., (*e*) Khalibit W.L.S., (*f*) West Chindwara W.L.S., (*g*) Gontara W.L.S., (*h*) Rukhand W.L.S., (*i*) Rangwan dam W.L.S., (*j*) Chachai Falls W.L.S., (*k*) Kawardha W.L.S., (*l*) Langur-Kharia W.L.S., (*m*) Turur-Baria W.L.S., (*n*) Balai W.L.S., (*o*) Amarkantak W.L.S., (*p*) Gonapalli W.L.S., (*q*) Sainmura W.L.S. and (*r*) Hasdo Basen W.L.S.

MAHARASHTRA

(*a*) Radhanagar N.P./W.L.S. (*b*) Nagzira N.P./W.L.S. (*c*) Wood Fossil N.P., (*d*) Bhamragarh N.P./W.L.S., (*e*) Kolhapur W.L.S., (*f*) Koyna N.P./W.L.S., (*g*) Mahabaleswar W.L.S., (*h*) Rev. Fr. Santapan W.L.S., (*i*) Itiadoh Rajoli W.L.S., (*j*) Gnangana N.P./W.L.S., (*k*) Tipeshwar W.L.S., (*l*) Darekasa W.L.S., (*m*) Mayani Lake W.L.S., (*n*) Nandgar W.L.S., (*o*) Ujani W.L.S., (*p*) Akola Grassland W.L.S., (*q*) Wadali Black back W.L.S., (*r*) Lalling Dhulia W.L.S., (*s*) Vikroholi W.L.S., (*t*) Dasgaon W.L.S. and (*u*) Achra W.L.S.

MANIPUR

(*a*) Dzuko W.L.S., (*b*) Dzuko N.P., (*c*) Siroi W.L.S., (*d*) Siroi N.P., (*e*) Imphal Botanic garden W.L.S., (*f*) Kaihlam W.L.S. and (*g*) Taret Lakhao W.L.S.

MEGHALAYA

(*a*) Tura Arabella N.P. (*b*) Rongrengri W.L.S., (*c*) Nongkhlaw N.P., (*d*) Shillong Peak N.P., (*e*) Nawsmai W.L.S., (*f*) Garampani W.L.S. and (*g*) Saipung Link W.L.S.

MIZORAM

(*a*) Dampa W.L.S., (*b*) Dampa N.P. (*c*) Rengdil W.L.S., (*d*) Twai W.L.S., (*e*) Ngengpui W.L.S., (*f*) Phawngpui N.P., (*g*) Palak W.L.S., (*h*) Murlen N.P. and (*i*) Murlen W.L.S.

NAGALAND

(*a*) Intanki N.P./W.L.S., (*b*) Kisa W.L.S., (*c*) Rangapahar W.L.S., (*d*) Dzuko-Pulicbadze W.L.S., (*e*) Shiloi W.L.S., (*f*) Macaque W.L.S.

ORISSA

(*a*) Billar Kanika N.P., (*b*) Gahirmatha North W.L.S., (*c*) Bhetnoi W.L.S., (*d*) Mahendragiri W.L.S., (*e*) Drirampur W.L.S., (*f*) Belghat W.L.S., (*g*) Chandrapur W.L.S., (*h*) Satkosia Gorge N.P., (*i*) Malyagiri W.L.S., (*j*) Nanda Island N.P. and (*k*) Chakapadkandon W.L.S.

PUNJAB

(*a*) Harike N.P., (*b*) Siali Dhar W.L.S., (*c*) Nava W.L.S., (*d*) Dholbaha W.L.S. and (*e*) Siswan Dulran W.L.S.

RAJASTHAN

(*a*) Desert N.P., (*b*) Kanodwala Rann W.L.S., (*c*) Siwana W.L.S., (*d*) Jodhpur Safari W.L.S., (*e*) Salvas W.L.S., (*f*) Ramgarh W.L.S., (*g*) Wood Fossil N.P., (*h*) Diyatra-Bap W.L.S., (*i*) Nokha W.L.S., (*j*) Tal Chhappar W.L.S., (*k*) Panch-padra W.L.S., (*l*) Sariska N.P., (*m*) Mount Anu N.P., (*n*) Kumbhalgarh N.P., (*o*) Sitamar N.P., (*p*) Alwar Fort N.P., (*q*) Johdia, Alwar W.L.S., (*r*) Sajandab W.L.S., (*s*) Bassi W.L.S. and (*t*) Boroswar W.L.S.

SIKKIM

(*a*) Dzongri W.L.S., (*b*) Tolung W.L.S., (*c*) Nimphu W.L.S., (*d*) Khang Chendzonga N.P. (extn.), (*e*) Kitam W.L.S. and (*f*) Pangola N.P.

TAMIL NADU

(*a*) Paint Calimere N.P., (*b*) Gulf of Manar N.P., (*c*) Mundan thurai -- Kalakad N.P., (*d*) Mundanthurai -- Kalakud W.L.S., (*e*) Megamalai W.L.S., (*f*) Anamalai W.L.S., (*g*) Nilgiri Tahr N.P., (*h*) Mudamalai N.P., (*i*) Koonthakuhum W.L.S., (*j*) Karangulum W.L.S., (*k*) Fichavaram W.L.S., (*l*) Kazhiveli W.L.S., (*m*) Marakemem W.L.S., (*n*) Karikili W.L.S., (*o*) Sandal W.L.S., (*p*) Kaveri W.L.S., (*q*) Sujjalkuttai W.L.S., (*r*) Boluvampatti W.L.S., (*s*) Barbetta W.L.S., (*t*) Karandamalai -- Sirumalai W.L.S., (*u*) Varus hanad W.L.S., (*v*) Palni W.L.S. and (*w*) Keenparai W.L.S.

TRIPURA

(*a*) Trishna N.P., (*b*) Central catchment W.L.S. and (*c*) Rawa W.L.S.

UTTAR PRADESH

(*a*) Pindari W.L.S., (*b*) Govind Pashu Vihar N.P., (*c*) Kedarnath N.P., (*d*) Yamunolri W.L.S., (*e*) Dodital W.L.S., (*f*) Dudhatayoli W.L.S., (*g*) Chakrala Deodhar W.L.S., (*h*) Banog W.L.S., (*i*) Ranikhet Pine W.L.S., (*j*) Nainital Oak W.L.S., (*k*) Binsar W.L.S., (*l*) Rajaji N.P., (*m*) Lal valley W.L.S., (*n*) SohelwaW.L.S., (*o*) Sohgibarwa W.L.S., (*p*) East UP Wet Grassland W.L.S., (*q*) Pipri W.L.S., (*r*) Dry grassland w and (*s*) Faisabad W.L.S.

WETS BENGAL

(*a*) Jaldapara N.P., (*b*) Buxa (extn.) W.L.S., (*c*) Badamtam W.L.S., (*d*) Mal 13 W.L.S., (*e*) Birik -- Rongpo W.L.S., (*f*) Dalka W.L.S., (*g*) Raiganj W.L.S., (*h*) Tista W.L.S., (*i*) Soharai W.L.S., (*j*) Ashidoh W.L.S., (*k*) Adhsoi W.L.S., (*l*) Singabad W.L.S., (*m*) Halna W.L.S., (*n*) Gompur W.L.S. and (*o*) Ajodhya W.L.S.

SECTION C

A WISDOM TO CONSERVE NATURAL RESOURCES OF ANDAMAN AND NICOBAR ISLANDS

13

ANDAMAN AND NICOBAR ISLANDS HAVE WIDE BIOGEOGRAPHICAL QUALITIES AND ARE RICH CONSERVATION ZONES

-- The Andaman and Nicobar Islands are two distinct groups of islands geologically, geomorphologically, biologically and biogeographically.

-- The Islands are unconsolidated marine sediments composed and mud clay and sand with only sandstone, mud stone and shale as harder substratum material. There is no crystelline rock ; weathering and erosion is, therefore, high; The sandy or crystalline substratum offer easy percolation and little surface run-off, whereas in muddy and clayey substratum percolation is less and run-off is high. Increase in turbidity in Continental shelf area damages planktonic seagrass flora and fauna including coral formation. There is little ground water.

-- The relict of three ancient tribes viz., the Sentinelese, the Jarwas and the Onges are invaluable wealth of the islands. While the negrito tribes inhabit the Andaman Groups, the Nicobar groups have inhabitant of mongoloid group.

-- The assemblages of various floral and faunal elements (Indian, Burmese, Indochinese, Japanese, Chinese, Malayan, Indonesian and Indo-Pacific) circumscribed in a limited area give the island a unique status for detailed biogeographical studies Fig. 13.1.

-- There is high degree of endemism among plants (more than 10%), avifauna (more than 39.2%), Insectivora (all the six

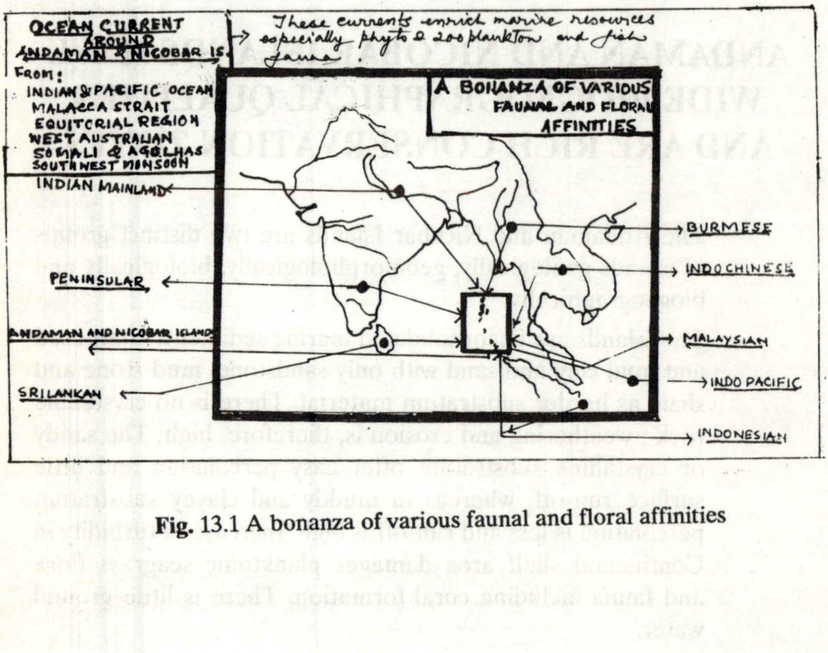

Fig. 13.1 A bonanza of various faunal and floral affinities

species are endemic i.e. 100%), Chiroptera (46.1%), Rodenta (more than 73.3%) and several other examples may be presented. This gives the islands a distinct position comparable to Galapogus island in pacific ocean and the vulnerability of species for confinement over a limited space.

-- Many animals are confined exclusively to each group of islands viz. the Andaman group and The Nicobar group, some however, occur in both the island groups.

-- Carnivora is poorly represented and the poor mammalian population is an indicator of islands a distinctive character.

-- The islands have five of the seven species of Teatudines of the world.

-- In the Order Squamata (reptiles other than turtles and a species of crocodile) endemism is as high as 32 per cent.

-- The islands have at least 16 species of mammals, amphibians, reptiles and birds which are rare and endangered.

-- The islands are rich in coral fauna, fish fauna (more than 700 species as per Sir F. Day F.R.S. puts it).

-- The resources of crustacea, mollusc, prawn, lobster, edible oyster, mussel, black-lip pearl oyster, clam, turbo and troctus, sea cucumber and sea urchin, marine turtle and other maricullune potential.

-- Dipterocarps, the dominant and indicator genus of the tropical forests of Southeast Asia, occurs in Andaman group only (not in Nicobar group) while the Nicobar group has Tree palms and many species common to Malaysia, Indonesia and other southeast Asian Islands. The plant association has influenced animal species and their association.

-- The islands have many species of plants, animals and birds that do not occur in Indian mainland, but occur in Burma, Malaysia, Indonesia and in other South-east Asian countries.

-- Owing to large extension of continental shelf (34900 sq.km.) the islands have a wide resources of planktonic (Zoo and phyto) flora, sea weeds, corals and other animal resources.

-- Various types of ocean currents (some from Pacific Ocean

and some from Antarctica) reach the Bay of Bengal and Andaman Sea, that bring with them various zooplanktonic and phytoplanktonic fauna and flora respectively to this region. The physical chemical and biological characters of the marine environment are therefore, changing continuously. Fig. 13.1.

-- The islands are cyclone prone and the maintenance of vegetation cover is essential. Long periods of geographical isolation of these islands from the adjacent land masses and subsequent changes in the environment have largely contributed to the evolution of many of their endemic species and sub-species.

-- The speciation in these islands has also encouraged the rich variety of isolated biotopes, favouring unchecked evolution of a variety of species.

-- The distribution of higher vertebratb groups of terrestrial animals such as reptiles, birds and mammals in the islands throws valuable light on the true insular characteristics and the extent of morphological variation due to geographical isolation.

-- Coral reefs of the fringing type are available in abundance on the eastern side, of the islands while barrier reefs occur on western side.

-- The Andaman fauna have greater affinities with that of Indo-Chinese sub-region, while the Nicobar fauna shows more Indo-Malayan elements in the south.

-- These islands have higher percentage of endemic species distributed over a small area and are always threatened from the pressure of human activities. Insular biology has special status and conservation area for a threatened species should be set up with much more areas than a threatened mainland species as "Smaller the land area the higher rate of normal extinction". Mc Arthur and Wilson (1967) in their work entitled "The Theory of Island Biogeography (Princeton University Press, N.Y.) have established the fact that the smaller the land area the higher the rate of natural extinction".

14

THE ISLANDS AND FORESTS

Background Information

The Andaman and Nicobar Islands consist of a chain of 348 large and small islands and islets situated in bay of Bengal. The tract lies between the latitude 13°-41° and 6° North and the longitude of 92° and 93°-7° East. The main island runs almost in a chain parallel to North-south direction towards southeast of India. There are 324 islands in Andaman group and 24 in Nicobar group and they extend over 754 km. north-south.

There are two distinct island groups in the archipelago, the Andaman group north of ten degree Channel and the Nicobar group south of it Barren island (erupted on 30.4.91) and Narcondon Island are of volcanic origin while the others are the tips of a submarine mountains extending from Sumatra (Indonesia) to Aracans in Burma.

The highest point in the archipela go is Saddle peak in North Andaman rising to a height of 732 m above sea level. The coastal strip is narrow.

The islands have a population of 1,88,254 (1981 Census). Of this the aboriginals account for less than 1000.

The total geographical area covered is 8293 sq.km. of which forest area is 7171 sq.km. Only 34 out of 348 islands are inhabited.

Physiography, Geology, Soil and Climate

The terrain of these islands in irregular and is more of less hilly to undulating, sometimes flat in the valleys. The hills following the general direction of the islands, lie nearly north and south. Spurs and ridges run-off the main ranges in all directions, resulting in a very confusing topography. Slopes are moderate to steep, aspect is towards east and west. The geographical configuration of Andaman Islands is described to be the southern continuation of Arkan-Yoma tectonic unit. Geological formation of North, Middle, South and Little Andamans is mainly with intrusion of basic and ultra-basic igneous rocks. The underlying rocks are essentially sandstone. Serpentine rocks are found in the higher altitude. The islands are eocenic sediments of sandstone growing into conglomerates. Andamans are a part of old Gondwana land mass, and are of very ancient in origin.

The soil is mainly acidic in nature. Soil texture is medium on the surface and heavy in the deeper layer. Humus and organic matter deposition in the soil is very poor. Scarcity of sweet water is a constant problem due to the soil texture and humus content. Preservation of rain water in the rainy season is not possible and that caused erosion and due to lack of conservation of soil moisture in the dry season the soil becomes devoid of moisture.

Andaman groups of Islands lie in the tropical wet zone. Climate is warm and humid. The season can be divided into summer and rainy. The extreme summer and winter is particularly unknown. Maximum mean temperature range is from 26° C. to 31° C and the range of mean minimum temperature is 19° C to 23° C. Mean annual rainfall during last four years was 3240 mm at Mayabundar, 3158 mm at long Island and 2920 mm at Portblair. Humidity is very high for most part of the year.

Thick eocene sediments deposited on pre-tertiary fine grained sandstones, shales, salt stones and limestones with instrusion of basic and ultra basic igneous rocks. Sand stones and commonest rocks, porous and slightly micaceous. In hill top there is green and red jaspers, quartzite crystalline limestone, occasionally gneissic outcrops. Car Nicobar is formed of grey clay, standstones and upheaved corals. In Nancowrie and Kamorta the bed rock is serpentine, covered by whitish or yellowish clay. Katchal is composed of calcarious sandstones. Little Nicobar and Great Nicobar has micaceous sandstones.

In Little Andaman mean annual rainfall varies from 3140 mm to 3200 mm. Mean annual temperature varies from 26° C to 27° C and maximum temperature rises to about 31° C to 32° C during April-May.

The Nicobar group of islands are surrounded by coral self and shallow seas of varying width, long and narrow stitch of sandy beaches. Car Nicobar and Katchal are almost flat ; Great and Little Nicobar are generally hilly, highest elevation being 700 m.

Soil depth in valleys is only 60 cm to 90 cm, soil is porus except in clayey and limestone areas. There is shortage of water in Feb. to April. Humus or organic matter is lacking.

Rainfall varies from 2750 mm to 4550 mm. There is no perennial stream except in Great Nicobar, Little Andaman and Katchal.

Legal Status

Prior to enactment of Indian Forest Act, 1927 which came into force in the year 1936 the entire forest of Andaman islands were administered under the provision of Andaman administrative legislation III of 1876. As per legal activities set up of the Andaman forest the land is entirely owned by the State and all land in Andaman is Government owned.

Reserved and protected forest areas are as follows :

	Reserved Forests	Protected Forests	Total
Andaman Group	2363 sq.km.	2571 sq.km.	4924 sq.km.
Nicobar Group	706 sq. km.	1541 sq. km.	2247 sq. km.
	3069 sq.km.	4112 sq.km.	7171 sq.km.

Out of this, 7 per cent (approx.) of the area is unfit for any use being too small and rocky ; 13 per cent form coastal belt which is to be preserved for protection against the soil erosion from sea and rain, 35.5 per cent earmarked for Aboriginal Tribes and 14 per cent under revenue settlement. The remaining area of 30.50 per cent is available for commercial forests.

Forests (Qualitative assessment)

The forests are evergreen, lofty, dense, 45 m or even more high, characterised by the large number of species of trees which occur together. Gregarious dominants (Consociation) are rarely met with and generally two-third or more of upper canopy trees are of species individually contributing more than 1 per cent of the total number, a few species, however, do occur semi-gregariously. Some species of the top canopy trees are with clear boles 30 m. long and 5 m. or more in girth and may be briefly deciduous without affecting the evergreen nature of the forest as a whole. The canopy is extremely dense, epiphytes are numerous, especially aroids, ferns, mosses and orchids. Climbers are very conspicuous in semi-evergreen and moist deciduous forests than in evergreen. Ground vegetation is not dense. The under growth is often a tangle of canes, creeping bamboos and palms, plank buttress is common, the leaves are thick and glossy. Cauliflory is common. Dipterocarps are very characteristic trees in the forests of Andaman (not in Nicobar islands) and therefore these forests form western fringes of humid tropic where Dipterocarps predominate. Besides palms and canes which are so conspicuous the bamboo *Dinochloa andamanica* and *Oxytenanthera nigrociliata* occur abundantly.

The following types of forests and some principal species as classified by Champion and Seth are found in the Islands. The edaphic and climatic factors play dominant role in vegetative formations.

Forest type	*Floristic composition*	*Type of land*
(i) Tidal Swamp forests (Mangrove forests) 4B/TS 2	Bruguiera conjugata, Rhizophora *mucronata, R. candelaria, Bruguiera parviflora, Avicennia officinalis, Nipa fruticans, phoenix paludosa, Sonneratia caseolaria, Areca triandra.*	Low lying banks of creeks and sheltered portions of the coastal line.
(ii) Little forests (Beach forests) 4A/li	I. *Casuarina equisetifolia, Manilkara littoralis, Bombax, insignis, Afzelia bijuga Terminalia catappa, T. bialata.* II. *Barringtonia asiatica, Erythrina variegata, Calophyllum inophyllum.*	Alluvial high level *along the sea* coast beyond the reach of the sea

Forest type	*Floristic composition*	*Type of land*
	III. *Pongamia pinnata, Thespesia populnea. Hibiscus tiliaceus, Hernandia peltata.* III.(a) *Pandanus tictorias, Phoenix paludosa, Heritiera littoralis.* IV. *Ipomoea pescaprae, Vigna retusa.* V. *Colabrine asiatica, Ixora* species, *Cycas rumphii.*	
(iii) Andaman Moist Deciduous Forests (3A/ci)	I. *Pterocarpus dalbergiodes Terminalia, manii, T. procera, Bombax insignis, Pterocymbium tictorium Dipterocarpus pilosus. Tetrameles mudiflora, Albizzia lebbek, Legerstroemia hypoleuca* II. *Dillenia pentagyna, Semicarpus* species *Cinnamomum* species, *Pterospermum acerioides, Lannea coromandelica, Cratoxylon formosum* II(a) *Bombusa lineata* III. *Murraya paniculata, Ixora gradifolia, Clorodendrum viscosum, Glycosmis arborea* IV. *Calamus* species, *Entada phaseoloides, Harrisonia brownii*	Shallow sandy or clay loam soils formed from the hard coarse grained sandstone hands of shales & Conglomerate and is found on hills upto and elevation of 100 m.
(iv) Andamans semi-evergreen forests. (2A/ci)	I. *Dipterocarpus alatus, D. pilosus, Terminalia bialata, T. procera, Bombax insignis, Artocarpus chaplasha, Albizzia lebbek, Pterocymbium tinctorium, Sideroxylon langepetialatum, Tetrameles nudiflora*	Main valleys on well drained immature alluvial soil, sufficiently old & raised above the level of salamty

Forest type	Floristic composition	Type of land
	II. *Myristica* species, *Pometia pinnata, Pisonia excelsa, Dillenia pentagyna, Xanthophyllum andamanicum, Talauma andamanica*	
	III. *Clerodendrum viscosum, Leea indica, Saprosma ternatum*	
	IV. *Dinochloa andamanica, Calamus andamanicus, C. palustris, Korthalsia laciniosa, Combretum extensum, Plecospermum andamanica*	
(v) Southern hill top tropical evergreen forests (1A/c2)	*Dipterocarpus alatus, D. grandiflorus, D. gracillis, Artocarpus chaplasha, Planchonia andamanica Hopea odorata, Baccaurea sapida, Endospermum chinese*	Tops of the high hills and sometimes a steep slopes lower down on poorer soils
(vi) Andaman tropical evergreen forests (1A/C2)	*Diptrocarpus costatus, Mesua ferrea, Canarium manii, Hopea andamanica, Cratoxylon formosum, Euphorbia trigona, Cryptocarya ferrarse, Phoenix* spp.	
(vii) Giant evergreen forest (1A/ci)	*Dipterocarpus alatus, D. grandiflorus, Artocarpus chaplasha, A. gomeziana Plachonia, andamanica, Sideroxylon longepetialatum, Pisonia umbellifera, Andamanthera pavonina, Hopea odorata*	Throughtout the islands alongside the banks of larger streams where soils are deep alluvial

Forest-resources -- Some Quantitative Assessment

Tree density

This study carried out by Forest Survey of India presents a picture of density of tree crop and their volume which is very high and, therefore, relevant in the present context.

(*i*) *North Andaman* : Stem density is very high i.e. 355-386 stem per ha.; about 61.6 per cent of trees are below 20 cm. diameter and only 10.2 per cent of the trees are above 10 cm. dia. Occurrence of higher diameter class is maximum in case of *Dipterocarpus* species followed by *Pterocymbium tinctorium*, *Artocarpus chaplasha* and *Terminalia procera.*

(*ii*) *South and Middle Andamans* : It is observed that the trees density in both the divisions is very high *i.e.*, 355.117/ha and 390.121/ha for Middle and South Andaman respectively. About 62.30 per cent of stems in Middle and 57.87 per cent in the South are below 20 cm diameter and only 3.48 per cent and 3.64 per cent of the stems are above 60 cm diameter. Regarding density of commercial species, *Dipterocarpus* species in South Andaman occupies the top place followed by *Parishia insignis.* In Middle Andaman, occurrence of *Dipterocarpus* is predominant followed by *Diospyrus marmorata.*

(*iii*) *Little Andaman* : In the Evergreen strata *Myristica* species occur in abundance with a density of about 11.4 stems per ha., but in volume estimation Gourjan (*Dipterocarpus* species) with a density of 5.8 stems/ha ranks the top with 23.560 m^3/ha.

In the littoral forests sea Mohua (*Manilkara littoralis*) is abundant with a density of 46.04 stems/ha and volume 115.8 m^3/ha while in deciduous forests *Dipterocarpus* is on the higher side with 9.1 stem/ha and volume 34.45 m^3/ha.

Distribution of Mangrove Species

A fair idea of density of trees in mangrove swamps can be had from the following :

The distribution of stems/hectare for various mangrove species

Name of species	North Andaman	South Andaman	Middle Andaman	Little Andaman
Rhizophora candelaria	800.6	756.0	317.698	Not assessed
Bruguiera conjugata	427.1	128.0	308.556	
Bruguiera parviflora	1.9	–	–	
Avicennia officinalis	148.9	–	–	
Heriterera littoralis	4.4	156.0	45.712	
Miscellaneous	48.00	100.0	62.854	

Total Growing Stock

A fair idea of biomass of tree can be had from the following study.

(*i*) *Little Andaman* : In view of the fact that substantial number of trees are mature to over mature and need early conversion a rotation of 75 years has been fixed in management schedule. Papita, Didu, white Dhup which are matchwood timber could be worked at a rotation of even 30 to 40 years.
Total growing stock has been calculated to be 17517.645 (000 m^3).

(*ii*) *North Andaman* : Total growing stock has been calculated to be 33,776.433 (000 m^3)

(*iii*) *Little Andaman and South Andaman* : The total growing stock in the middle and South Andaman Divisions are 13,839.849 and 13,977.532 respectively (both in 000 m^3).

Land utilization status of Andaman and Nicobar Islands

Areas in hectare)

Total geographical area	–	829300
Area under forest administration	–	714400
Area not available for cultivation	–	14431.11
Other cultivated land excluding fallow land	–	40820.97
Fallow land	–	3576.16
Total cropped area	–	14953.48
Net area sown	–	14040.71
Area sown more than once	–	917.77

Source : Andaman and Nicobar Statistical Bureau).

15

ANDAMAN AND NICOBAR ISLANDS ARE BIOGEOGRAPHICAL BONANZA

Definition

Various biogeographers have defined biogeography in their own language. Pierre Banserean defines : "Biogeography covers origin, distribution, adaptation and association of plants and animals, their solution in changing environment and also involves recent and ancient geological history of the area." Others think that man's impact on the landscape, landuse pattern, urbanization, industry, browsing and grazing pattern, fire cycle, lumbering auteecology and population structure, structure of vegetation, dynamics of vegetation, analytical phytosociology etc. cover the definition. Udvardy M.D. puts, ethology, ecology, animal ecology, systematics, synecology, auteecology, pedology evolution, geology, phytogeography and evan embryology and areography under the definitions of zoogeography.

Zoogeographical Classification

P.L. Sclater (1858) classified zoogeography based on bird distribution. But later Illiger (1815), Swainsen (1835) made attempts to catalogue faunas. Woodward (1851) used the phylum Molusca and based 27 zoogeographical regions on them. Schmarda (1853) characterised his 21 land "realms" by various leading species or higher Taxa a mammals, birds and insects largely corresponding to the sub-regions of Wallace (1876). Sclater's zoo-geographical system was widely recognised but not influenced by the Darwin-Wallace theory. However, detailed discussion on this subject, is not being done here. In 1876, Alfred Russel Wallace published "The Geographical Distribution of Animals' in which he supported Sclater's regional system as preferable as to that of Allen, a view with which most of his contemporaries agreed. Wallace emphasised historic and evolutionary factors and also climates, vegetation, geological history etc. into consideration.

Wallace in 1976 divided oriental region into (*i*) Indian subregion (*ii*) Himalayan or Indo-Chinese subregion and (*iii*) Indo-Malayan or Malayan sub-region ; the Andaman and Nicobar Islands falls under the sub-region. In 1888 Blanford proposed the divisions: 1. Tibetan 2. Himalayan 3. Indian 4. Malabar or Ceylonese 5. Burmese and 6. South Tenasserim. Sharp however proposed divisions of India into (*i*) Indian peninsular Sub-region (*ii*) Indo-Malayan Sub-region (*iii*) Indo-Chinese Sub-region (*iv*) Himalo-Malayan Sub-region and (*v*) Himalo-Chinese Sub-region.

Talbot (1939) recognised 8 major divisions of India (on the basis of distribution of Lepidoptera mainly Rhopalocera) viz. Ceylon, Peninsular India, Northwest India, West Himalayas, Northeast India, Burma, Andaman and Nicobar Island, Talbot's Northeast India shows affinity partly to the Palaearctic and partly to the Oriental, with infiltration from the South.

Alcock (1910) recognised six territories in the distributional pattern of fresh water Crustacea, viz., the western territory, the western Himalayan territory, the northeastern frontier or territories, Burma-Malay territory, peninsular territory and Indo-gangetic plain territory.

Annandale (1911), based his division on the distribution of fresh water sponges polyzoa (this confirmed Alcock's division).

Christopher (1933) made his division on the distribution pattern of Culicidae viz., Trans-Indus area, the Indo-gangetic area, the peninsular area, Malabar-ceylon area, Assam-Burma area, and Himalayan area.

Malcolm. A. Smith (1931, 1935, 1941) recognised that the reptiles of Thailand, Indo-China, South China, Burma, Assam and the Eastern Himalaya are very closely related (a single zone). The fauna of this vast zone is very closely related to that of the Malabar tract of western ghats within the peninsula. He recognised the sub-regions-- 1. Desert area 2. Kashmir & Himalayas 3. Gangetic plain 4. Central India 5. The Deccan 6. Malabar mountain & Ceylon 7. Chota

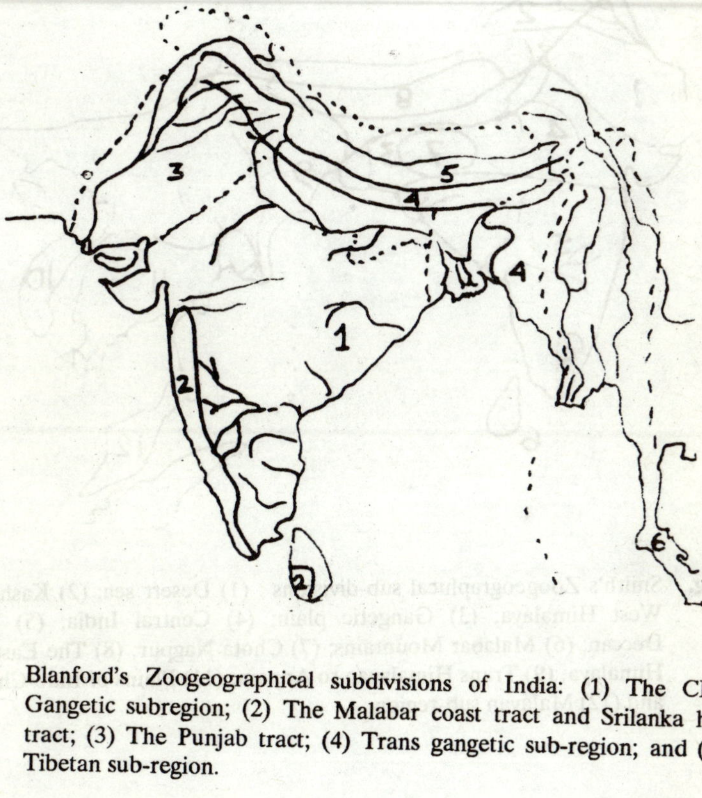

Fig. Blanford's Zoogeographical subdivisions of India: (1) The CIS-Gangetic subregion; (2) The Malabar coast tract and Srilanka hill tract; (3) The Punjab tract; (4) Trans gangetic sub-region; and (5) Tibetan sub-region.

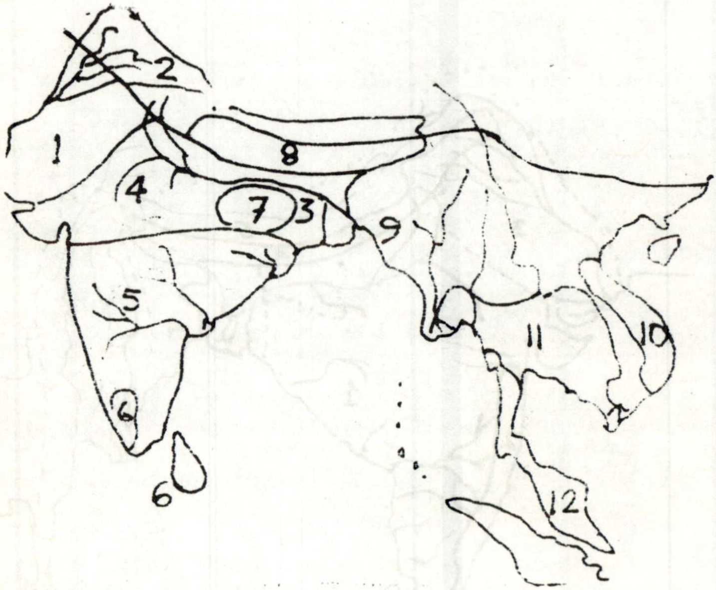

Fig. Smith's Zoogeographical sub-divisions : (1) Desert sea; (2) Kashmir, West Himalaya; (3) Gangetic plain; (4) Central India; (5) The Deccan; (6) Malabar Mountains; (7) Chota Nagpur; (8) The Eastern Himalaya; (9) Trans Himalayan to Annam; (11) Plains of Indo-China; and (12) Malayan sub-region.

Nagpur. 8. Eastern Himalayas 9. Trans Himalayan 10. Anam 11. Plains of Indo-China and 12. Malayan Sub-region.

On the pattern of distribution of land mollusca, Blanford (1870) divided India into four major regions viz., the Punjab, India, East Bengal, Malabar with Srilanka. Indian region was sub-divided into several sub-regions. In 1876 Blanford pointed out the presence of African element in Indian fauna.

In 1873, Elwes considered India and Malaya peninsula as a single geographic unit which he named Indo-Malayan Region (the Himalayan or the Himalo-Chinese sub-region, the Indian sub-region and the Malayan sub-region).

Various Biogeographical Realm, Region and Kingdom: (a brief Summary) :

A. Sclater's realm.	I.	Palaeogean
		1. Palaearctic 2. Aethiopean 3. Indian 4. Australian 5. Nearctic.
	II.	Neogean
		5. Nearctic 6. Neotropical regions.
B. Huxley's realm	I.	Arstogaea
		1. Circumpolar region 2. Notogaea.
C. Allen's realm		1. Arctic 2. northern temperate 3. American tropic 4. Indo-African 5. South American temperate. 6. African temperate 7. Australian 8. Antarctic.
D. Wallace's region.		Almost like Scalater's classification except 3 which is Oriental.
E. Engler's kingdom		1. Northern extra tropical 2. Paleo Tropical 3. South America 4. Old oceanic.
F. Blandford's realm	I.	Arctogaea
	II.	Notogaea
	III.	Neogaea.

There are other conceptions too. Mention must be made of Spate D.H.K. who in his work "India and Pakistan -- A general and regional geography" (1957) has summarized various classifications and put India under four broad groups viz., A. The mountain rim, B. The Indo-gangetic plains, C. The peninsula, D. The Islands. He puts Arunachal Pradesh (under A) -- Eastern Himalayan tagging Bhutan with it; Andaman and Nicobar has been put under the group D. Some biogeographhers put the group D under modified Udvardy's classification.

Himalayan Uplift and Biogeographical Dynamic

The biogeographical composition, distribution and the Himalayan uplift patterns are interlinked. Original Gondwana flora and fauna in both composition and general character were influenced and modified by the uplift of the Himalayas. It was primarily through the formation of biogeographical routes of interchange and barriers with Eurasia, and to a limited extent with parts of Africa. The Himalayan uplift established a direct connection between the ancient Peninsular Block of India and Asia in the north-east, whereas, the Peninsular block was rammed against the young Tethyan sediments of Asia forcing these sediments to flow around in the form of the Himalayan and Burmese Arcs.

After the disappearance of the Tethys sea the north-east and the older peninsular block came into contact with the young rock giving rise to the north Burma-Northeast Indian region. This region is the confluence of flora and fauna of the southeast Asia which migrated west and south-west India. This also helped migration of South and Southeast Chinese, Indo-Chinese and Malayan elements in India. Arunachal Pradesh, Nagaland and Manipur regions may be called floristic and faunistic gateway of India. The area holds and pivotal place in the historic process of progressive evolution of the present day flora and fauna.

The faunal and floral movement in the Himalayas forest had been from the east to the west. In peninsular region movements had been from the east and also from the west. The north-western route opened only where the entire Tethys had been completely obliterated ; then migration took place from Europe, Africa and other places.

The Arunachal-Assam-Nagaland region is the western-most boundary of the range of many Indo-Chinese mammals, like certain squirrels and the easternmost limit of the distribution of many peninsular species such as the spotted deer.

The mammals of the islands Assam, Nagaland are primarily Indo-Chinese, though there is admixture of the Peninsular and Ethiopia elements ; the palaearctic montane elements also intrude in it. Several relict species of peninsular India (in western ghats) are also distributed in the island.

Lacunae in Classification

In Bio-geographical Division, the zoologists and the botanists have not jointly worked to collate various theories. Even the zoologists have postulated their own theories (Malaclost, Herpetologist, Dipterist, Lepidopterologist, Ichthyologist, Ornithologist and others). The proposed theories of phytogeographical (based on vegetation and climatic relationship) and zoogeographical divisions of India depict only partial picture of the present day distributional patterns.

Vegetation structure is changing and so the faunal structure and composition ; they interact with each other and are indivisible components of nature and constitute an indivisible whole. So it may be summed up " The ecology and Biogeography of India are thus essentially part of a Complex mosaic and the events in India are an integral phase of the larger biogeographical and geomorphological evolutions and involves, therefore, a number of problems of great fundamental importance, based on ideas of Continental Drift which climaxed in the uplift of the Himalayas and cannot be studied in isolation". Unfortunately such importance has not been felt so far.

16

PHYTOGEOGRAPHY

Phytogeographical Bonanza

(*a*) Andaman and Nicobar Islands have 2270 species of Angiospern and 150 species of Pteridophytes of which 220 species are endemic. Endemism is rich in specific level (not generic). Rubiaceae has 23 species and 2 genera endemic in island. Among non endemic flora 50 per cent belong to India ; Other 50 per cent are distributed to South East Asia, Burma, Thailand, Malaya and Indonesia. Of the 220 species (endemic) recorded over the year, it was difficult to locate at least 135 species; have they been extinct?

(*b*) About the rare and endangered plant species of Andaman and Nicobar Islands, the Botanical Survey of India has marked 123 species as rare and endangered. Some of the predominant families are Anonaceae, Palmae, Orchidaceae, Melastomaceae, Guttiferae, Euphorbiaceae, Rubiaceae : Others species belong to another 56 families. Of the Gymnosperm and Pteridophytes mention may be made of family Gnetaceae, Ophioglossaceae, Podocarpaceae and Psilotaceae.

(*c*) Rare and Endangered nonendemic species of islands which occurs in South-east Asia number about 136. Principle families being Orchidaceae (17 species) Rubiaceae (14 species), Euphorbiaceae (11 species) Anonaceae (6 species) Cyperaceae (6 species).

(*d*) there are 19 Monotypic genera in these islands which show how endangered these genera are.

(*e*) Great Nicobar, which has no *Dipterocarpus* or *Pterocarpus* species, is very rich in various species of tropical forests. Many species have been found endemic to the island. Some endemic species are *Cyrtendra burttii, C. occidentalis, Cyrtandromoea nicoborica, Jainia nicobarica, Ophiorrhiza nicobarica, Macaranga nicobarica, Ophiorrhiza infoundibularis, Phrynium paniculatum, and Phanera nicobarica.*

(*f*) In great Nicobar about 158 or about 20 per cent of species has distribution in South-east Asia and not found in India. Some are : *Dicranopteris curranii, Microsorium pleropus, Pteris griffithii, Vittaria elogata* var. *angustifolia, Syngramma alismefolia, Crepidomanes bilabiatum, Vandenboschia maxima, Lindsaea chienii.* etc.

(*g*) In great Nicobar itself there are 33 species of orchids belonging to 26 genera. Orchid flora show close relation with Indonesia rather then Burma. In fact there are common genera of orchids of Thailand, Burma, Indo-China, Malayasia, Indonesia, Philippines and Sri Lanka.

(*h*) Many genera of Nicobar Islands such as *Cyathia, Astronea, Otanthera, Cirtandra, Stauranthera, Kibara, Stemonurus* and some other species do not occur in Andamans. The flora of these tropical forests are highly niche specific, spaciation is being continued all along. The unique flora trapped and confined to these islands are now being subjected to large scale biotic pressure.

Some New Taxa Recently recorded

Anoectochilus nicobaricus - (Nicobar Island), *Artabotrys nicobariacus* - (Nicobar Island), *Kurziodendron* sp., - (Andaman Island), *Eulophia nicobarica* - (Car Nicobar), *Glochidion airyshawii* - (Andaman), *Phanera nicobarica* - (Great Nicobar Island), *Jasminum cordifolium* - (Andaman), *Jasminum multiflorum* Var. *nicobaricum* - (Nicobar Island), *Alyxia reinwardtii* var. *lucida* - (Andaman), *Bulbophylhum lilacinum* - (Andaman), *Coelogyne thailandica* - (Andaman), *Cymbidium pubescens* - (Nicobar Island), *Dendrobium indragiriense* - (Andaman), *D. plicatile* - (Nicobar), *Erycibe griffithii* -

(Andaman & Nicobar), *Macaranga triloba,* (Great Nicobar), *Mussaenda villosa* - (Great Nicobar) *Neodlissochaeta celebica* - (Great Nicobar), *Nervilia punclata* - (Great Nicobar), *Phraetia secunda* - (Andaman), *Pteroceras alatum* - (Nicobar), Pteroceras alatum - (Nicobar), Spathoglottis plicata - (Nicobar), *Sterculis macropada* - (Great Nicobar Island), *Tino mischium petiolare* - (Great Nicobar Island), *Vrydaqzynea albida* - (Great Nicobar Island), *Cyrtandra burttii* - (Great Nicobar Island), *C. occidentalis* - (Great Nicobar Island).

New phytogeopraphical record:

Diplospora adnamanica, Hypolytrum balakrishnii, Jainia nicobarica, Jasminum andamanicum, J. unifoliolatum, Macaranga nicobarica, Malleola andamanica, Memecylon collinum, ophiorrhiza infundibularis, O. nicobarica, phanera nicobarica, Phrynium paniculatum, Phyllanthus andamanicus, Taeniophyllum andamanicum, Tinomiscium nicobaricum, Vernonia andamanica, Zingiber maximum, Tadehaqi triquetrum, Cresnine Javonica, Glochdion calocarpum and Trigonostemon viridissimus.

17

ZOOGEOGRAPHY

Smith (1931, in Fauna of British India) was of opinion that the Andaman Island contain an impoverished Burmese fauna; the Nicobar Islands approximate to the Sumatran type. Reptilean fauna indicate Malayan type.

Endemicity of some land and litloral animals:

	Group	No. of species	No. of endemics	Percentage
Land	Mammals	46	32	69.56
	Birds	242	95	39.25
	Reptiles	72	24	33.33
	Amphibians	18	6	33.33
	Spiders	62	28	45.16
	Moths and Butterflies	366	86	23.56
	Dragonflies	36	7	19.44
	Termites	26	14	53.84
	Antlions	7	3	42.85
	Earthworms	28	8	28.57
Littoral	Reptiles	11	0	0
	Fishes	560	24	4.28
	Holothurians	68	12	17.64
	Chitons	13	1	7.69
	Hermitcarb	38	2	5.26
	Copepods	172	48	27.90
	Amphipod	110	8	7.27
	Polychaetes	186	8	4.30
	Archinnelids	18	4	22.22
	Nematodes	54	6	11.11

Endemicity is some mammals :

– *Shrew (Order Insectivora)*

Of the six species, 3 belongs to Andaman and 3 to Nicobar Islands. All of these species are endemic. The species are – Andaman island spiny shrew, Millers Andaman Spiny shrew, Nicobar spiny shrew, Jenkins Andaman spiny shrew and Nicobar tree shrew.

– *Bats (Order Chiroptera)*

Of the reported 32 species, 12 species are endemic (6 species each of Andaman and Nicobar Islands).

Ecobiological studies of such animals in tropical evergreen forest may open vast field of science.

The species are Narcondom small flying fox, Car Nicobar flying fox, Nicobar lesser short-nosed fruit bat, Andaman lesser short-nosed fruit bat, Nicobar pipisterelle and several others.

Other Chiroptera of Andaman (A) and Nicobar (N) Island groups are :

(A) Andaman Flying Fox, Malayan large flying fox, Indian flying fox, Short-nosed fruit bat, Dobson's long tongued fruit bat, Blackbeared tomb bat, Malay fats vampire, Tickell's bat, Blyth's clubfooted bat.

(N) Malayan large flying fox, Indian flying fox, Blyth's pouch bearing bat, Fulvus leafnosed bat, lesser yellow bat, bent winged bat.

– *Macaque (Order Primates)*

Pig tailed and Nicobar crab eating macaques are recorded. The latter is endemic and the former is of doubtful occurrence.

– *Carnivora (Order)*

Andaman masked palm civet is endemic. The occurrence of jungle cat is doubtful.

– *Pigs (Order Artiodactyla)*

Andaman wild pig and Nicobar wild pigs are both endemic.

– *Deer (Order Artiodactyla)*

Barking deer and spotted deer are available which have been introduced.

– *Rats and Squirrels (Order Rodenta)*

Of the 14 species of rats, 13 species are endemic (6 in Nicobar Island group). The species are – *Rattus rattus andamanensis, Ratus rattus artidorsum, Rattus rattus flabilis, Rattus rattus holchu.*

Others in Andamans (A) and Nicobar (N) groups exclusively are : *Rattus rogersi* (A), *R. stoicus* (A), *R. palmarum* (A), *Rattus pulliventer* (N), *Rattus burrus (N), Rattus burrulus* (N), *Rattus burrescens* (N), *Mus musculus* (A).

– *Marine Mammals*

Dugong, common dolphin and false killer are the species of which last two are endangered.

Avifaunal Resources

The Bombay Natural History Society had made a thorough investigation of avifaunal resources of Andaman and Nicobar Islands. They have identified 255 species. They were of the opinion that except for the Megapode and Nicobar pigeon, the avifaunal affinities appear to be closure to India than to Burma and Malaya. 94 species are endemic.

Endemic species (a few) are :

Andaman green heron, Nicobar Tiger bittern, Andaman grey teal, Andaman black crested, baza, Car Nicobar shikra, Katchal, shikra, Andaman crested hawk eagle, Andaman pale serpent eagle, Nicobar crested serpent eagle, Great Nicobar crested serpent eagle. Andaman dark serpent eagle, North Nicobar Megapode, Nicobar blue breasted quail, Andaman blue breasted banded rail, Andaman banded crake, Andaman white breasted water hen.

Dove and Pigeon – High degree of Endemism in

Andaman grey fronted green pigeon, Nicobar green imperial pigeon, Andaman green imperial pigeon, Pied imperial pigeon,

Andaman wood pigeon, Nicobar coockoo dove, Andaman coockoo dove, Burmese red turtle dove, Andaman emerald dove, Nicobar emerald dove, Nicobar pigeon.

Endemism in Parakeet

Large Andaman parakeet, Andaman red breasted parakeet, Blyth's Nicobar parakeet, Andaman red-cheeked parakeet, Nicobar red-cheeked parakeet.

Endemism in Owls

Andaman barn owl, Andaman scops owl, Andaman lesser scop owl, Nicobar lesser scops owl, Hume's brown hawk owl, Andaman brown owl, Nicobar brown hawk owl.

Endemic kingfisher

Andaman blue eared kingfisher, Andaman three-toad forest kingfisher, Andaman stork billed kingfisher, Nicobar stork billed kingfisher, Andaman ruddy kingfisher, Andaman white breasted kingfisher, Andaman white collared kingfisher, Nicobar white collared kingfisher.

Others – Andaman chestnut headed bee-eater, Andaman broad billed roller Narcondom hornbill, Andaman black woodpecker, Andaman fulvous breasted pied woodpecker, Nicobar hooded or green breasted pitta, Andaman black naped oriole, Nicobar blacknaped oirole, Andaman black headed oriole, large Andaman drongo, small Andaman drongo, Andaman racket tailed drongo, Nicobar drongo, similar 30 other species may be cited.

Endemic sunbird

Andaman olive backed sunbird, Nicobar olive backed sunbird, Car Nicobar olive backed sunbird, Nicobar yellow backed sunbird.

The following are winter visitors :

Common teal, Horse feeld's Goshawk, Asiatic Sparrow-hawk, Pole harrier, Maopagu's Harrier, Marsh harrier, Osprey, Eastern

Peregrine falcon, European kastrel, East Himalayan kastrel, Eastern Baillon's crane, Collared pratincole, Geey plover, Eastern golden plover, Large sand plover, European lille ringed plover, Pamir's lesser sand polver, Whimbrel, Eastern whimbrel, Eastern curlew, Bartailed godwit, Common red shank, Green shank, Green sand piper, Wood sand piper, Terek sand piper, Common sand piper, Turnstone, Pintail snipe, Fantail snipe, Great snipe, Jack snipe, Eastern k bot, Eastern little stint, Little stint, Temminck's stint, Longtoed stint, Curlew sand piper, Broad-billed sand piper, White winged black tern, Javan gull billed tern, Emerald Cuckoo, White cheeked Grey Drongo, Rosy pastor, Olive flycatchar, Brown flycatcher, Eastern red-breasted flycatcher, Northern bluethroat, there are also 10 species of warblers, species of Wagtail, 3 species of pipets and 2 species of Bunting

Reptiles

Of the 80 species of reptiles and amphibia known to the islands, 31 species visit sea shore at some stage of their life (indicates necessity of protecting sea shore). Stoloczka, thinks that several species of lizards and snakes (common to both the island groups) and other animals resemble Indo-Malayan, gradually passing into Burmese. Some of the species are :

Snakes (Order Squamata)

40 species have been recorded of which brown pit viper, Andaman pit viper, Colubrine Amphibious Sea Snake, Andaman banded Krait Boie's Cat Snake, Nicobar water snake, Brown wolf snake, Biswas's wolf snake, Andaman bronze snake, Nicobar stripedneck snake, Oat's blind snake, Brown spotted pit viper, Cantor's pit viper, Andaman banded kukri and Andaman blink snake are endemic.

Other Reptiles (Order Squamata)

Of 33 species the following are endemic (10 species) : Curltailed gecko, Andaman pay gecko, Green forest lizard, Tiwari's garden lizard, Andaman garden lizard, Andaman skink, Tytler's skink, Nicobar tree skink, white striped skink and Andaman water monitor.

Amphibia

There are 8 species of frogs and 2 species of toads of which Pillai's Narromouth frog and Nicobar toad are endemic. Others are brown frog, paddy frog, Andaman paddy frog, Nicobar frog, salt water frog, Boulenger's narrow mouthed frog, tree frog, Indian toad and Nicobar toad.

Rare and endangered species

Andaman wild pig, crabeating macaque, dugong, Estuarine crocodile, Green sea turtle, Hawkbill turtle, Water monitor, Andaman teal, Hawks, Large falcons, Narcondom hornbills, Nicobar megapode, Nicobar pigeon, Osprey or fish eating engle, White breasted sea eagle.

Insect fauna

Much information is not available about insect fauna of the island. Some records available in the records of Zoological Survey of India are summarised below :

Odonata	–	A.R. Lahiri (1975 Rec. ZSI); 31 species in South Andamans of which 5 are endemic, 3 common to India, 4 common to Sri Lanka and 10 to South-east Asia.
	–	Coleoptera – 19 species of aquatic coleoptera (aquatic beetles) have been identified.
Termites	–	Roonwal and Bose (Rec. ZSI 1964) identified 17 species of which 10 are endemic, 6 species are common with oriental region. (Genera – Neotermes, Cryptotermes, Prorhinotermes Schedorhinotermes, Hospitalitermes, Nausetermes, Odontotermes).
Spiders	–	B.K. Tikadar (Rec. ZSI 1977) records 41 genera and 58 species of spiders. The work involved also those of Rhorrel and Pockok.

Fish

Hora, identified (in 1938) 112 species of fish in 40 families. But Sir Francis Day, FRS estimated about 700 fish species out of which 372 species were identified. Major genera of fishes identified were Tunies, Perches, Sharks, Corangids, Sardines, Makeral etc. Tuna shoals are often found within 5^0 North and South of Equator. Campbel bay is 6^0 4' North lat. has this fish in between November and May. Nicobar Island has annual potentiality of Toon fish of 50,000 tonnes.

Coral

It will be an exhaustive list to mention various species of corals that occur in the area. Of the vast coral resources it may be mentioned that exploration and identification is far from complete at this stage. The following are common genera which need be mentioned :

Acropora, Montipora, Pocillopora, Staylophora, Pavona, Euphyllis, Symphyllia, Merulina, Favidae, Goniastrea, Porites, Goniopora, Fungia etc. which have more than 150 species in Andaman water.

Crustacea

Several genera of crabs are of importance which are *Scylla, Uca, Sesarma, Matuta, Graspeus, Calappa, Eriphia* etc. Of the hermit crab *Clibanarius* has several species. More than 400 species of Crustacea have been recorded in Andaman water.

Mollusc

Many species of Conus have been collected from this locality. Of the mollusc, about 800 species have so far been recorded from Andaman and Nicobar areas. However, *Nautilus, Cassis, Turbo, Trochus, Tridaena, Lambis, Mure, Nerita, Trocus, Telescopium, Lyncina* may be mentioned of the area.

Other invertebrate animals :

Sponges	–	70	species recorded
Coelenterates	–	117	species recorded
Stony corals	–	179	species recorded
Polychaetes	–	161	species recorded
Crustaceans	–	407	species recorded
Pycnogonids	–	8	species recorded
Sipuncalid	–	20	species recorded
Echinoderms	–	331	species recorded

General

(*i*) The Africal Land Snail (*Achanita fulico*) introduced in Portblair has now become a menace in some islands.

(*ii*) The mollusc Oreobba represents only Nicobar Islands, while *Haughtonia* the Andamans. The cerembycid beetle *Artimpanza* is also known from Sumatra and Borneo. The island has a number of endemic elements, Indochinese and Malayasian forms and a few Peninsular relicts. The species which are peculiar to the Andaman and Nicobar Islands include snakes *Typhlops, Lipeltis woodmasoni* etc. The other endemic elements of the animals are the earthworms *Pheretima osmastoni, P. suclores, laptoxena* sp., *Graphium agamemnom* (a lepidoptera of Malayasian origin) and *Agamemonon* sp., etc. The Assam-Murmis butterfly *Papilio memnon* extends to the South-Nicobar Island. The Indo-Chinese and Malayasian elements of the vertebrates are exemplified by *Xenopettis unicolor* in the Andamans. Many other examples may be cited.

The indigenous landfauna have not been thoroughly identified. The Nematode, earthworm, leeches, isopod, crab, harmit crab, centipede, scorpion, spider grastropods, frog toad, lizard, snake, tortoise, rats, bat etc. have not been properly collected and identified. Similar is the situation with beetles, bugs, butterflies, dragonflies, grashoppers, thrips, mosquitoes, flies, bees, wasps, ants and whiteants which form bulk of the insect fauna.

Fresh water fauna has also not been properly investigated. The marine fauna is well represented in the inter tidal zone (sandy, muddy and rocky beaches, algal weeds and mangrove) coral reefs, plankton, necton and benthos.

18

MARINE AND COASTAL RESOURCES

Primary Production in Coastal Water

In east Bay of Bengal the lower boundary of the photosynthetic layer in between 84--99 m and the Continental Shelf has high rate of production. There is seasonal fluctuation of the layer. International Indian Ocean Expedition (I.I.O.E.) has been collecting extensive data on the continental shelf of Indian coast. Mariculture activities in the coastal areas from Diglipur to Campbell Bay provide some information in the marine resources.

The temperature of coastal areas ranged from 27-32°C and salinity 27--33 per cent. The pH ranged around 7.85 in the mangrove areas.

A study of Phytoplankton reveals that of 56 species of phytoplankton identified. 22 species belong to Diatomaceae, 33 species to *Dinophyceae* especially the species of *Ceratium, Peridimium, Ornithocercus, Dinophyris, Phalacroma, Pyrocystis* and *Pyrofacus*; near the shore occurs the species of Diatomaceae besides *Rhizosolenia* sp., *Guinardia flaccida, Biddulphia sinensis, Streptothea indica* and *Chaetoceras* sp. The blue green algal *Trichodesmium thiabantii* has been collected from Nicobar islands. Near Nicobar Dinoflegellates dominate. Quantity of phytoplankton was as high as 6100 cells/m³ in the Andaman sea compared to the low values (1200 cells/m³ in Arabian sea). A higher phytoplankton production (5000 cells/m³) was found to the South West to Nicobar islands.

Chlorophyll values differ seasonally. There is higher

phytoplankton production near the Nicobar island (300-500 u9/m³), Little Andaman and part of South Andaman.

The occurrence of these algae and their abundance are directly related to highly productive animal life.

Ecological Status of Shore

A cross-section of coastal sites are presented below. The rich and varied faunal resources are indicated by varied ecological sites. (**Source** : CMFRI, Bull. 34, Sept., 1983)

Andaman Islands

	Table Islands	-	Rocky shore
	Digliepur	-	Muddy creak
	Mayabandar	-	Rocky, Sandy boulder sandy
	Shoal Bay	-	Muddy
	Hovelocks	-	Rocky
Port Blair	North Bay	-	Reef
Port Blair	Navy Bay	-	Muddy shallow
Port Blair	Phoenix Bay	-	Slushy mangrove shore
	Sesostris Bay	-	Muddy
	Ross island	-	Fouled reef
	South point	-	Rocky reef
	Corbyn's Cove	-	Rocky reef
	South Burma nalla	-	Sandstone flat
	Wandoor	-	Slushy mangrove
	Chiriyatappu	-	Sandy reef
	Ruth land blard	-	Rocky area in deep water
	North Cinque	-	Sandy
	Hut bay	-	Sandy
	Butler bay	-	Rock inter tidal sandy

Nicobar Islands

	{ Hog punt Sawai, Passa }		Sandy & rocky
	Treetop falls		
Car Nicobar	Malacca	-	Reef
	Kimios back water	-	Marshy mangrove
	Arong	-	Rocky
	Katchal		
	East Bay	-	Sandy reef
	West Bay	-	Rocky
	Camorta		
	Jetty -		Marshy reef
	Kanaka	-	Sandy reef
	Nan Cowry (Jetty)	-	Sandy calm
	Spiteul bay	-	Calm mangrove gringed
	Mangrove	-	Murshy
	Trinkat Is.	-	Thick reef
	Great Nicobar--		
	Campbell bay	-	Sandy with dead coral
	Vijaynagar	-	Sandy reef

Zooplankton in Coastal Water

Only selected areas have been investigated which do not reveal the richness of resources of planktonic flora and fauna.

Abundance of Copepode, Calamaid Copepod, molluscan larvae, bulk occurrence of larvae of penaeid prawn have been marked in some sites; molluscan larvae, Chaetognaths, decapod larvae lucifer and appendicularian are interesting fauna. Pteropods noted in Shoal bay, Lucifer in Camorta, Gastrpod larvae in abundance noted near Nicobar Island and some special sites in coastal water.

Coral Reefs and their Environment

Coral formation in Andaman and Nicobar Islands extends from 92°-30° to 94° E. longitufe and 7° to 14° N. latitude. About 135 species (in 59 genera) have been identified of which 110 species of 45 genera are hermatypes and 25 species of 14 genera are hermatypes. The zonations of hermatypes in this area has not been studied in detail. The two major forms are the massive forms and the ramose forms each with a set of reef building and reef swelling species. The massive corals of this area, as in Indo-pacific regions, are *Favin, Favites, Platygyra, Sympphyllea, Goniastrea, Diploastrea* and *Porites*. The branching forms are *Pocillopora, Stylophora, Seriatopora, Aeropora* and *Montipora*. There is a foliaceous forms like *Montipora foliosa, Echinopora lamellosa, E. horrida* and *Pavona* sp.

The dead and living corals in the subtidal zones are excellent living substrata for molluscs, crustaceans, sponges and many worms *Drupa, Pyrene* and *Cerithium* are gastropods that are found among the branches of ramose corals. Pearl cysters live attached to the top and sides of ramose and massive corals.

Many bivalnes like *Acra* and borer like *Lithophaga* are in plenty on dead and living corals.

Mangrove Resources

Mangroves of Andaman Island occupy an area of 115200 ha mostly along the creaks. They occur at the fringes of littoral and evergreen forests – the width of these forests would be just 100--300 m. There is a very long coastline where mangrove thrives. The mangrove swamps protect an abundance of microscopic and microscopic algae. The higher forms are *Padina gymnospora, Laurencia, Papillosa, Dictyota dichotoma* and *Ulva lactuca* and micro algal forms are *Enteromorpha, Chaetomorpha, Cladophora, Ectocarpus* etc.

An association dominated by seagrasses *Zostera, Thalassia, Cumodocea* can be seen in bay of Nan Cowry and Konaka region of Camorta Island. The mangrove of the creeks have various species of Seweeds of animals such as *Trochus, Turbo, Cerithedium, Tlescopium* edible oyster, Crab, *Ocypode, Periophthalmus, Ambasssis, Therapon, Penaeus, Metapenaeus, Scylla, Portunus* etc.

Resources of mangrove have also been discussed in this chapter.

Sea Weed Resources

These are found primarily in intertidal zone on the rocky coasts.

55 species of Sea weeds were collected in Andaman and Nicobar Islands (16 species of Chlophyceae, 17 species of Phaeophyceae and 22 species of Rhodophyceae,. The distribution of some species are as under :

Place		Genera/Species
Diglipur	-	*Turbinaria, Sargassum*
Mayabandar	-	*Padania, Turbinaria, Sargassum Dictyota, Hormophysa*
Little Andaman	-	*Laurencia, Gracillaria, Halimeda*
Car Nicobar	-	*Dictyota, Turbivaria, Hormophysa, Gracilaria*
Kalchall	-	*Amphiroa, Galaxuara, Turbivaria, Sargassum*
Camorta	-	*Ulva, Halimeda, Laurencia, Gracilaria*
Nan Cowry	-	*Sargassun, Turbinaria, Padina*
Great Nicobar-	-	*Gracilaria, Turbinaria*
(Campbell bay)	-	*Turbinaria, Gracilaria, Ulva*

Penaeid Prawn Resources

19 species in 6 genera of such fish have been identified *Penaeus merguiensis* is the dominant species accounting for 49 per cent, floowed by M. dobsoni 42 per cent. The creeks and protected bays of mangrove swamp and the high tidal amplitude offer best breeding fround.

Lobster Resources

12 species of lobster have been identified. There are ample facilities for developing these resources.

Crab Resources

Out of species of edible crab *Scylla serrata, Portunus pelagicus* and *P. sanguinolentus* are exploited. The first two species are cultivated in

India.

Edible Oyster Resources

Andaman and Nicobar Islands have *Crassostrea madrasensis* and *Saccostrea cucullata* which are edible and are found in intertidal and subtidal areas. The population has been found to be maximum in North Andaman Coast gradually decreasing in Nicobar groups. The percentage of *C. madrasensis* is less than *S. cuculata.* The culture of these Oysters is possible in sheltered bays and creaks.

Mussel Resources

Green mussel, *Perna viridis* attach themselves to rocky substratum and in the roots of mangroves. In 1 sq meter area 14 specimens may occur. The sheltered bay and creeks offer breeding and rearing sites.

The Black-lip Pearl Oyster Resources

The indo-Australian region is rich in Pearl Oyster. The Mergui acchipelango which is the eastern limit of the Andaman Sea is the area of Burmese pearl culture with the most valued species Pinetada maxima. The distribution of pearl Oyster, though recorded from the Andamans, is not known. The marine ecosystem of the Andaman and Nicobar Islands with numerous bays has been considered potential grounds for pearl culture. The Andaman species of Oyster are *Pinctada margaritifera, P. fucata, P. sugillata, P. anomioides*. The Oyster grow on hard stoney substraction. the occurrence of P. *margaritifera* over many island coasts have been recorded. Further investigation is necessary to record occurrence of other species and their density. It is very likely that gold-lip or silver-lip pearl Oyster P. *maxima* occurs in Andaman and Nicobar Islands. *Pteria penguin* – a black-winged pearl oyster has been collected from Mayabandar, Havelock island and Camorta. The species are used to produce brilliant pinkish half-pearls in Fifi and Borneo islands.

Giant Clam (Tridacna) Resources

Tridacna crocea, T. maxima and *T. squamosa* are giant clams and live in shallow water of the coral reef fromations. *Tridacna* serve as hosts to a number of other organisms such as shrimps as commensels. Four of the six Indopacific species occur in Andaman and Nicobar Islands. The occurrence of the species in intertidal and subtidal areas has been rendered from Diglipur to Campbell bay. These species are exploited for meat and for shell.

Trochus and Turbo Resources

Trochus niloticus and *Turbo marmoratus* are two species of mollusc which occur in abundance in the Andaman and Nicobar Islands and are of much economic value. The former has been recorded from all the islands. Both the species are edible. The present annual landings of *Trochus* from Andaman island vary between 400-500 tonnes and of *Turbo* 100-150 tonnes. The market value of *Trochus* is Rs. 4000/- per tonne; *Turbo* shell cost Rs. 10,000/- per tonne. Middle, South Andamans and among Nicobar group of islands Terssa, Trinkat, Katchall, Nar Cowry and Campbell bay of Great Nicobar are rich areas for these shells, Barren island is also rich in *Turbo* species.

Sea Cucumber and Sea Urchin Resources

In the sheltered bay and lagoons these animals breed and subsist happily. The sea cucumbear prefers muddy or sandy flats, the sea urchins rocky costs and algal beds.

Of the 40 species of sea cucumbear found in the Andaman and Nicobar Islands, only a few species are useful as commercial species.

Holothuria atra, a sea cucumbear, is abundant in the islands and ocean on dead coral reef with sandy and muddy patches. It prefers areas where calcarious alga *Halimeda* sp. to feed on *H. scabra* occurs in abundance in Andamans. Besides. *Actinopyga mauritiana, A. echimites. A, mimiaries* and *A. seratidens* are also found in these islands are of commercial importance.

Sea Urchin Resources

The gonads of sea urchins are said to be a delicacy and are eaten fresh adding a little vinegar lime

Tripneustes gratilla is found on the algal bed. *Diadema secotosum* is found in the wall of Jetty at Aberdeen (Port Balir) and Nan Cowry harbour. *Echinometra mathaei* occurs on the rocky coasts.

The mud flats at Diglipur, Mayabandar, Janglighta near Port Blair and Campbell bay are favourable area of culture of sea cucumber. Of the eight species of sea urchin and sea cucumber, *Holothuria scabra* is ideally suited for culture.

The prospects for beche-de-mer industry in Andaman and Nicobar Islands is bright.

Sea Turtle Resources

Of the seven species of sea turtles found in the ocean of the world, four species *viz.,* Olive Ridely Turtle (*Lepidochelys olivacea*) Green turtle (*Chelonia mydas*), Hawkbill turtle (*Eremochelys imbricata*) and Leather back turtle (*Dermochelys coriacea*) occur in Andaman Islands.

Sea turtle meat and eggs constitute an important food source in many countries. Turtle soup prepared from cartilageneous tissue called "calipee" is widely relished. The horny laminate on the shell of the Hawkbill turtle are made into high valued curios. Carefully supervised "Turtle watching night" to observe the turtle behaviour, nest formation etc. has become popular to tourists in Australia and Malaysia.

Turtle nesting beaches have been destroyed by construction of structures on the Betapur Coast in Middle Andaman.

On a survey it has been found that Leatherback turtle occurs in abundance in little Andaman in Wets Bay and South Bay) than the other three species. At one time Wandoor became the largest turtle depot and butchering centre in South Andaman.

The turtles occur in Landfall island, Table island, Turtle island, South island, Long island, Strail island, Betapur shore, Ross island, Burmanaka, Chiritapu, Wandoor, Cinque island, Hutbay, Butter bay, South bay, Bompoka island, Katchall island and elsewhere.

Crocodile Resources

The saltwater Crocodile (*Crocodylus porosus*) is the world's largest reptile and produces most valuable crocodile leather. Muggar nullah and Kalpong river at Diglipur were formerly crocodiles den; also found in Landfall and North reef islands, Karala Creek, Balmi creek, Casharina bay, Hadson bay, Bakultala, Papitadera, Kadantulla and elsewhere. Even today West Coast of North Andaman and the

Kalighal – Parangare area has an estimated 250 to 350 numbers in North, middle and South Andamans. The forest department has not yet taken up any breeding programme of this species as in West Bengal, Orissa and some other states.

The economics calculated by R. Whitakar, may seem unrealistic, puts the value of each crocodile hatchling to Rs. 800.00 (for skin, meat fat and gall bladder) after 2 years 6 months growth. Net gain be calculated from each nest is Rs. 16,000/- per annum.

Mangrove Resources

Natures masterpieces of creation are primarily in the creeks in North, Middle and South Andaman Islands. Satellite imagery interpretation has revealed an area of 1,15,200 ha under mangrove forests mainly in the three Andaman Islands (north, middle and south).

Principle creeks are Buchanan bay creek, Elizabeth bay creek, Casuarina bay creek, Mepherson creek, Balmi creek, Moare bay creek, Beele bay creek, Gribb creek, Candle bay creek, Teroppa Balyo creek, Throughfan creek, Blair bay creek, Laximpur bay creek, Diglipur creek, Kerala creek, Jagannathcera creek, Austin strait creek, Betapur dina creek, Toliging creek, Barion zig creek, Porlogzig creek, Humpray strait creek, Yeratila zig creek, Needham creek, Showal bay creek, Zirakatum creek and several others where mangrove forests have luxuriant growth.

Coastal mangroves are unique resources that are also useful to man. Estuaries are complex biological sites which are rich in nutrients and micrbiological organism and other animate objects which are linked by process resulting in a continuous exchange and assimilation of energy. Part of the life cycle of many marine organisms are spent in the mangroves. Many species of Penaeied shrimps spawn off shore using mangrove as refuge and feeding ground. There is a positive correlation between shrimp landings and mangroves forests. Food web structure in coastal zone is extremely complex. Benthic micro algae and organic detritus (seal of algae micro organism constitute the principle energy sources for vast animal population. The mangrove support and sustain marine fisheries and as feeding ground for many marine invertebrates and fish by providing large amount of detritus for marine food. These forests serve as a link between terrestial and marine ecosystem which are highly productive. Floral and faunal sucessional studies are of unique scientific interest. Mariads of species and individuals invertebrates, their specialised life-cycle colouration and behaviour pattern in coastal and sandy shore are specific qualities of the mangroves. Shrimp and Oyster cultivation may give a yield of 17 tonnes per ha per year valued more than Rs. 3,78,000, mussle may yield 180 tonnes per ha per year. Sea weed may also be cultivated. Mangrove forests, capture fisheries and open water aquaculture are largely complimentary activities which are one form of land use, removal of mangroves forests affect capture and open aquaculture. This indirectly means also the disappearacne of mangrove dependent species. The highly productive and dynamic sites can be very usefully utilised by judicious management to uplift the standard of living of the poor people. The mangrove forests counteract the force of cyclonic wind; they also provide sites for recreational and aesthetic purposes.

The destruction of mangroves forests may eliminate very many species of fish shrimp and prawn, Acid sulphate that is produced in soil where tree felling has been done is indicative of misuse of mangrove sites. So all mangrove sites in both Andaman and Nicobar Islands must be protected.

19

WILDLIFE SANCTUARIES AND NATIONAL PARKS

Sl. No.	Name	Notification	Area in (ha)	Remarks
A. National Parks				
1.	Wandoor Marine	CF/WL/43-Vol. I dt. 24-5-83	28150	
2.	Mt. Harrist	Silva/G-88 dt. 13-11-79	4662	
3.	Saddle Park	–do–	3254	
4.	North Button	–do–	44	
5.	Middle Button	–do–	44	
6.	South Button	–do–	03	
B. Sanctuaries				
7.	Narcondom Is.	CF/WL/ 19 dt. 19-2-77	681	
8.	Barren Is.	–do–	161	
9.	North Reef Is.	–do–	348	
10.	Interview Is.	–do–	13387	
11.	South Sentinel Is.	–do–	16100	
12.	Baltimalve Is.	CF/Wl/50/Vol. 1 dt. 1-1-85	223	
13.	Tillanchong Is.	–do–	1683	

Sl. No.	Name	Notification	Area in (ha)	Remarks
14.	Megapode Is.	CF/Wl/50/Vol. 1 dt. 1-1-85	12	
15.	Loha barrack Crocodile Is.	G/118-Vol. i 'dt. 25-7-83	106	
16.	Bingham	CF/WL/50-Vol. I dt. 19-1-87	8	Entire Island
17.	Sir Hugh Rose	–do–	106	–do–
18.	Pitman	–do–	137	–do–
19.	James	–do–	210	–do–
20.	Potanma	–do–	16	–do–
21.	Kyd	–do–	800	–do–
22.	Patric	–do–	13	–do–
23.	Defence	–do–	1049	–do–
24.	Montogemery	–do–	21	–do–
25.	Clyde	–do–	54	–do–
26.	Sandy	–do–	154	–do–
27.	Snake	–do–	3	–do–
28.	Cinque	–do–	951	–do–
29.	Passage	–do–	62	–do–
30.	Sisters	–do–	36	–do–
31.	North Brother	–do–	76	–do–
32.	South Brother	–do–	124	–do–
33.	Bluff	–do–	114	–do–
34.	Spike	–do–	1170	–do–
35.	Talabaicha	–do–	321	–do–
36.	Mangrove	–do–	39	–do–
37.	Stoat	–do–	44	–do–
38.	Belle	–do–	8	–do–
39.	Arial	–do–	5	–do–
40.	East of Inglis	–do–	355	–do–
41.	Duncan	–do–	73	–do–
42.	Oyster	–do–	8	–do–
43.	Parkinson	–do–	34	–do–

Sl. No.	Name	Notification	Area in (ha)	Remarks
44.	Cone	CF/WL/50-Vol. I dt. 19.1.87	65	Entire island
45.	Land fall	–do–	2948	–do–
46.	East	–do–	611	–do–
47.	West	–do–	640	–do–
48.	Peacock	–do–	62	–do–
49.	White Cliff	–do–	47	–do–
50.	Reef	–do–	174	–do–
51.	Mayo	–do–	10	–do–
52.	Paget	–do–	736	–do–
53.	Shearme	–do–	785	–do–
54.	Ponit	–do–	307	–do–
55.	Ox	–do–	13	–do–
56.	Shark	–do–	60	–do–
57.	North	–do–	49	–do–
58.	Kwangtung	–do–	57	–do–
59.	Rowe	–do–	1	–do–
60.	Latouche	–do–	96	–do–
61.	Jungle	–do–	52	–do–
62.	Tribly	–do–	96	–do–
63.	Table (Excelsior)	–do–	169	–do–
64.	Table (Delgarno)	–do–	229	–do–
65.	Temple	–do–	104	–do–
66.	Turkle	–do–	39	–do–
67.	Ross	–do–	101	–do–
68.	Brush	–do–	23	–do–
69.	Bamboo	–do–	5	–do–
70.	Blister	–do–	26	–do–
71.	Dot	–do–	18	–do–
72.	Curlew	–do–	3	–do–
73.	Gliver	–do–	16	–do–
74.	Oyster	–do–	21	–do–
75.	Grchid	–do–	10	–do–

Sl. No.	Name	Notification	Area in (ha)	Remarks
76.	Curlew (B.P.)	CF/WL/50-Vol. I dt. 19.1.87	16	Entire island
77.	Egg	–do–	5	–do–
78.	Swamp	–do–	409	–do–
79.	Dottrel	–do–	13	–do–
80.	Gurjan	–do–	16	–do–
81.	Sea Serpent	–do–	78	–do–
82.	Snake	–do–	73	–do–
83.	Bondoville	–do–	255	–do–
84.	Buchanan	–do–	933	–do–
85.	Surat	–do–	31	–do–
86.	Entrance	–do–	96	–do–
87.	Benett	–do–	346	–do–
88.	Roper	–do–	146	–do–
89.	South Reef	–do–	117	–do–
90.	Mask	–do–	78	–do–
91.	Taft	–do–	29	–do–
92.	Hump	–do–	47	–do–
93.	Gander	–do–	5	–do–
94.	Goose	–do–	1	–do–
95.	Flat	–do–	936	–do–
96.	Spike	–do–	42	–do–
97.	Ranger	–do–	426	–do–
98.	Wharf	–do–	11	–do–
99.	Tree	–do–	3	–do–
100.	Chanel	–do–	13	–do–

20

DEVELOPMENT ACTIVITIES AND CONSERVATION PROBLEMS

So far the qualities and floral and faunal resources of the Islands have been presented in a very broad outline. A few such qualities and associated problems are :

(i) vast floral and faunal resources

(ii) interesting geographical situation

(iii) bonanza of biogeographical affinities

(iv) poor status of soil, devoid of nutrients and extremely poor in water retention capacity

(v) high degree of endemism of floral and faunal species which are endangered because of a very limited space for each

(vi) very rapid increase in population (from about 30,000 in 1951 to 1,88,000 in 1981) resulting in pressure of land

(vii) too many wood based industries (36 numbers utilizing a heavy quantity of timber for which heavy felling has been made

(viii) soil deterioration owing introduction of coconut and oil palm plantation

(ix) vast developmental activities involving many Government Departments, settlement of large number of people and creation of ever increasing infrastructural facilities

(x) settlement of unspecified number of fishermen

(xi) lack of field based research

In view of these (mentioned only partly) it is imperative to consider the following seriously for conservation of resources :

(*a*) Should there be a total moratorium of all large scale development projects and on further settlement of people as it is thought that the present population has exceeded the bearing capacity of soil and will destroy the biological resources ?

(*b*) Should there be a drastic reduction in the number of woodbased industries at least to a third immediately ? This will minimise the quantum of timber supply to these industries and obviously less felling of forests will be involved. The Forest and Plantation Development Corporation has recently started two match factories and one saw mill in Little Andaman and at Mayabandar.

(*c*) Should there be complete suspension of further felling of forests in Little Andamans for planting oil palm ?

(*d*) Should an island the forests of which have acquired a balance after millions of years isolation, be suddenly subject to massive modification ?

(*e*) Should the encroaches who are in large numbers be allowed to settle in the island or exterminated from the island ?

(*f*) Should the Department of Forests guarantee the primary agricultural requirements like soil protection, water availability and maintenance of agroclimate are being served by the Forests ?

(*g*) Should there be more Govt. and Non-Govt. offices, research centres, institutions etc. as those will invite more people require more fuelwood, more timber, more land and create other related problems ?

(*h*) Should there be a levish use of expensive timber in building construction ?

(*i*) Should the settlers still enjoy free gift of timber and other forest produce ?

(*j*) Should this island provide employment for persons coming from mainland ?

(*k*) Should the forest be managed on commercial line exclusively ? Why the management should not be mostly conservation oriented ?

These are intrinsic limiting factors in Andaman and Nicobar environment which will delimit the development plannings and prove such effort as unproductive. The forests of Andaman and Nicobar Islands are only the relict of 'Tropical Rain Forests' in India and there is already indication of species loss. Therefore, these forests are also to be preserved to provide aesthetic and recreational facilities for the millions Indian tourists and to shelter myriads of vertebrate and invertebrate animals besides plants, to explore for scientific and limited purpose of economy.

Since 1965 several expert groups engaged by the Govt. of India have visited the islands primarily for determining (*i*) suitability areas for settlement of refugees, (*ii*) integrated resources development, (*iii*) accelerated agricultural development. The problem of deforestation, however, appealed the then Prime Minister of India, Smt. Indira Gandhi and in her initiative a Multidisciplinary Team of officials visited the island in 1975. This team and another landuse committee set up later stressed for collection of data and preparation of land use map etc. but none stressed the basic and urgent need and problems of conservation. It was Mc Vean the I.U.C.N. expert who visited the island in 1976 recorded for the first time the serious loss of forest resources and exploitation beyond the carrying capacity of the land exacerbated by unchecked immigration. He thought the islands tropical forest, coastal and marine ecosystem could be saved, and suggested complete suspension of forest clearance. This was not accepted by the Govt. of India. As a follow up, the Govt. of India appointed a team of Administrators who visited the island and recommended (*i*) preparation of base industries; and (*ii*) creating of a Corporation *viz.* The Forest and Plantation Development Corporation.

It is paradoxical that the barring Dr. Mc Vean neither the foresters, nor other scientists nor the administrators, thought of taking drastic steps on conservation of fragile soil and vegetation cover.

None except Dr. Mc Vean raised his voice against settlement of people after clearing forests. By now more than 1150 sq. km has been diverted for non forestry purposes. Encroached area is 2784 ha. The Department of Forests suspect more encroachment in interior area which has remained undetected because of inadequacy of the protection personnel. Repatriates from Burma and Sri Lanka have been accommodated in addition to largescale settlement of erstwhile

East Pakistan refugees. In Little Andaman 2864.5 ha were marked for rehabilitation and 200 ha. for Nicobarise settlement and another 164 ha. for defence. In Great Nicobar 7055 ha. were marked for ex-servicemen and 1355 ha. for defence service. In 1980-81, 1463 fishermen were engaged in fishing and rehabilated 126 families. If marine fishes have to be exploited heavily and aquaculture to be developed, a vast population has to be rehabilitated. This indicates a danger to the island and, therefore, needs judicious handling.

Provision for land for future requirement of Defence and O.N.G.C. (which may expand enormously should oil be struck) has to be made. Besides these, there would be Administrative personnel, with enormous staff size, Forest Department Staff and skeleton staff of other essential Departments only. Unless the Govt. establishments are restricted to the minimum the population pressure and their infrastructural requirements will nullify all conservation effort.

21

FOREST MANAGEMENT AND CONSERVATION STRATEGY

Forest Management

It will be evident from the historical background of the Department that management was oriented towards exploitation of commercial species for earning good revenue and to obtain regeneration of a few commercial species only. Organisation structure had been developed accordingly. Owing to the pressure of 36 woodbased industries and various development work there has been clossal demand for timber and entire forestry machinery has been geared to cause the supply. The forest Department feels that due to lack of proper understanding on the availability of timber many woodbased industries were set up; now they feel there is limited specific quality timber. So long the rate of supply was subsidised. Their demand for more timber is increasing over the years (2.05 lakhs cum a year). The Forest Department apprehend such demand is a threat to the protection of forest in the island particularly in the vicinity of the existing hemlets of settlements. Extraction of timber using elephants, tractors and locomotives in the slopes have caused appreciable erosion.

Although the Forest Department has fixed up 4085 ha. or 1,85,000 cum as annual yield the experts other than the foresters think that over felling is being done because, (*i*) area assessment has no sound foundation; (*ii*) there is no growth data to rely upon and the fixation of rotation and conversion period is far from correct. Canes and other minor forest produce are being overexploited. The Forest

Department extracted 1,38,126 cum of timber during 1985-86, but wastage of logs and timber over was also high. One protective measure, however, adopted was to stop extraction by the Contractors agency.

These situations are to be adequately analysed and a more conservation oriented working plan drawn.

Natural Regeneration of 41,975 ha. has so far been achieved. Clear felling of forests was done in the past to raise Teak, Padauk and Pyinma plantations, but this method has been discontinued (plantation since raised over 14,860 ha.).

Many experts, other than the foresters, have severely criticised the method of selection felling and the entire canopy lifting operation for obtaining regeneration of commercial species. They thought soil got much sunlight to dry up; burning damaged regeneration and made area drier; seedling of evergreen species failed to germinate and then establish. So the subject is still dabatable. There is no cell in the Department of Forests to monitor the regeneration operation and evaluate the forestry and environmental aspects of natural regeneration process. Mc Vean found natural seedlings on forest floor but observed that litter and humus is lacking on forest floor.

At present the Department of Forests is said to have shifted emphasis from the productivity of commercial timber alone to forest conservation.

The mangroves have been exploited heavily because of high calorific value of wood as fuelwood. The present method of management prescribes felling in belts/patches. This may have to be discontinued. The present requirement as stipulated by the Forest Department is about 6150 cum per year specifically meant for Electricity Department for power production.

The plantation and Forest Development Corporation is working as lumbering Department diluting the functions of silviculture, protectionary and regulatory functions. Oil palm plantation has been raised over 1593 ha. The projected target is 2400 ha. But at the moment the raising of plantation has been raised in Sarthankhari in South Andaman and at Kalchal.

Conservation Strategy

The area of National Parks and Wild Life Sanctuaries has been planned to be increased to 2,000 sq km or about 24 per cent of the

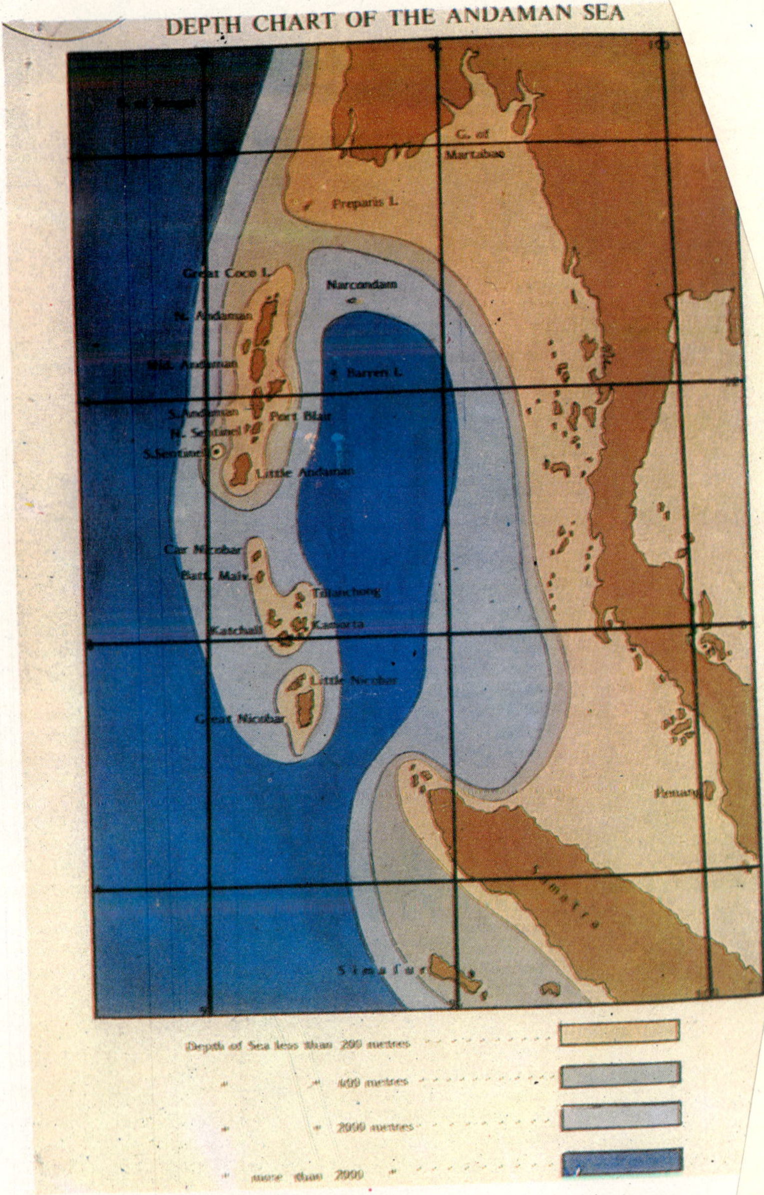

Andaman and Nicobar Islands in Bay of Bengal

geographical area of the territory. An area of 74,000 ha. has been notified to be Jarwa reserve and another 16100 ha as Senteneles reserve. The conservation status of this 79,990 ha. is not known. Tall trees might not have been damaged by these tribes, but what about small trees, shrubs, climbers and herbs etc. Yet it is certain that these areas will not suffer from biotic factors as other areas rehabilitated.

Analysing the fact presented in this and earlier chapters it is felt that conservation of small islands and other small areas would be inadequate to give full protection to the flora and fauna of this island in entirety. Conservation of total area of 2000 sq. km. of Sanctuaries and National Parks may initially satisfy many and boost up the percentage of National Parks and Wild Life Sanctuary areas in relation to geographical area but most of the plots are very small (except probably Interview Island) to give flora and fauna adequate space to interact, regenerate and servive as a Community and as individuals.

As such it is one further suggested that :

(*i*) Western half of Little Andaman island be converted into a Sanctuary.

(*ii*) Entire Little Nicobar (15902 ha.) island may be converted into a Sanctuary.

(*iii*) Northernmost part (covering about 20,000 ha.) of Great Nicobar island be converted into a Sanctuary.

(*iv*) All mangrove Forests be declared protected areas.

Forest Management vis-a-vis Environmental Change

The forests of Andaman are one of the richest repository of tropical forest resource in India. An area of about 1150 sq km. has been deforested since earliest settlement.

C.G. Rodgers, C.J. Lyal, Sir David Pran, CE Perkinson were the foresters and botanists that explored the forests and studied their resources. Regular plan was prepared by F.H. Todd in 1905 for North Andaman Forests which entailed improvement felling, adoption of strick fire control measures and seed dibbling in blanks. Boningtons working plan (1914) suggested felling of over mature trees. This was following by concentrated regeneration felling. Thereafter B.S. Chengappa (1952), J.C. Varmah (1960), ACD Cruz

(1970-71) working of the forests all over the north, middle and south Andamans.

In fact Padauk and Teak were raised at selected spots. By 1983, an area of 5046 acres was planted up but only a marginal success was achieved. In 1970's palm oil was introduced in Little Andaman by felling of virgin forests. Many consider the raising of teak plantation in Andaman as a resource mismanagement.

In a tropical evergreen forest micro-climate in the forests has to be preserved. Any alteration of it can only be counter-productive leading to forest degradation. Opening up the canopy beyond certain limit, allow more sunlight to penetrate, more heat created in forest floor and allow torrential rain drops to hit the soil destroying top soil and disbalancing nutrient cycle. Removal of a small percentage tree (as prescribed in Working Plans) created large gaps all over (as the trees had vast crown cover) and therefore humid atmosphere below top canopy changed. In felling operation, logging and extraction of timber trees countless regeneration of trees, shrubs and herbs were damaged. The movement of heavy vehicles, skidders and catterpillar tractors destroyed vast expanse of undergrowth and soil exposed to atmospheric acid especially to heavy rain. As a result hydrological characteristics of the worked areas changed vastly. Everywhere over felling of trees, damaged trees and cyclone affected trees were noticed especially in Little and North Andamans. The very nature of evergreen forests have changed due to felling of plywood and matchwood species over vast areas.

The foresters have already exploited the best areas as faras timber is concerned. It is difficult to find a glimpse of once existing giant tropical evergreen in the Andaman after walking scores of kilometres.

Geologically and geomorphologically in Andaman and Nicobar islands have certain special character which have a considerable bearing on human existence in there. These are natural climax forests. The islands are essentially unconsolidated marine sediments composed of mud clay and sand with only sand stone, mud stone or shale as harden substraction material. There is no crystalline rock and, therefore, weathering and erosion is extremely high. In sandy or coralline substation percolation is extremely rapid with little run-off while in muddy or clayey substation percolation is very little and run-off is very high.

Destruction of forests have increased the humidity of coastal water which might affect coral reefs in continental shelf.

The drainage in the hills is very rapid and the substratum can hardly retain much water. So there is dearth of ground water sources. The apparatus dampness and lashness of these islands were due to the unbroken climax forest which insulated the ground surface and retained humidity.

It has been an ecological disaster caused to Andaman forests by settling lakhs of agriculturists all over the islands. Forestry is yet to guarantee the protection of soil, water and maintenance of suitable agroclimate.

SECTION D

SIKKIM HIMALAYAS THE NUCLEUS OF NATURAL RESOURCES OF GREAT HIMALAYAN RANGE

22

CONSERVATION OF HIMALAYAN NATURAL RESOURCES IN SIKKIM –SINGALILA HILLS UNIQUE ROLE–

The Indian sub-continent is blessed with an immence variety of plant and resources which sustains millions of it's people. The preservation of floral and faunal genetic diversity and conservation of its species and entire ecosystem for sustainable utilization is of crucial importance for the future survival and development of our people. Various vital and invaluable habitats that sustain endemic flora and fauna have been depleted under the relentless human pressure. Many species have been extinct or threatened with extinction or are endangered. Forest areas have shrunk. The extermination of species, genera or depletion of ecosystem has resulted in permanent loss to science and disruption of the process of evolution. The diversity of our biological organisms is a vital resources which needs to be protected.

A World Conservation Strategy (WCS) was launched in India on March 6, 1980. The WCS sponsored the International Union for conservation of Nature and Natural Resources (IUCN) with the advice, cooperation and financial assistance of the United Nations Environment Programme (UNEP) and the World Wildlife Fund (WWF) and in collaboration with the Food and Agriculture Organisation of the United Nations (FAO) and the United Nations Educational, Scientific and Cultural Organisation (UNESCO) have endeavoured to safeguard the destruction of earths natural resources. Various countries including ours have already responded to its suggestions.

It is evident, therefore, that there is compelling conservation need

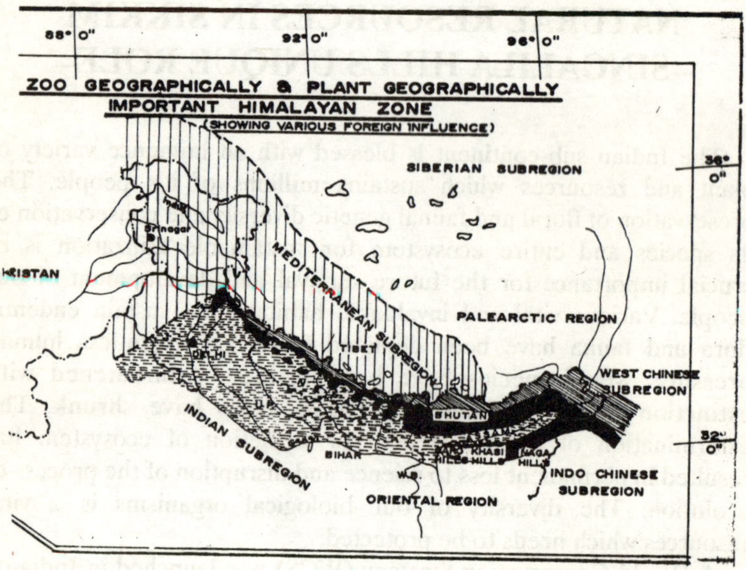

Fig. 22.1: Zoogeographically and Plant geographically important Himalayan zone.

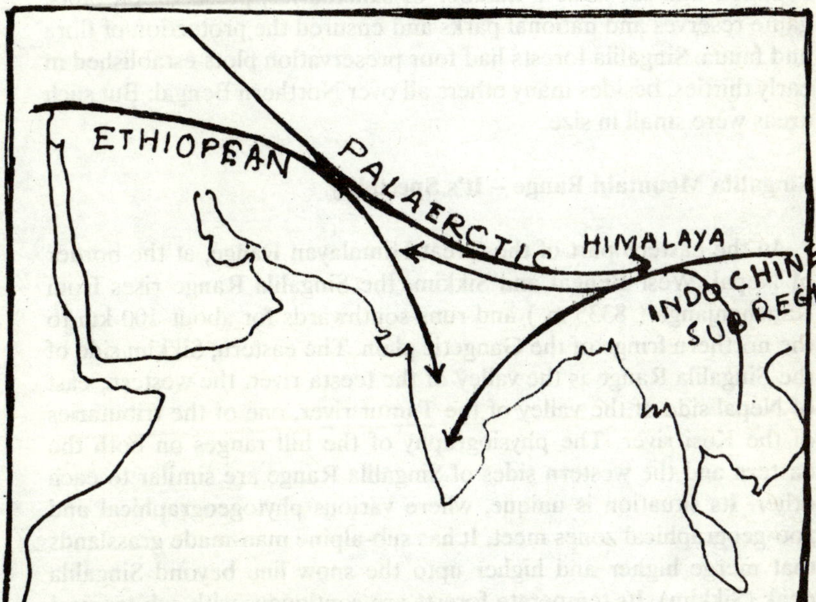

Fig. 22.2 The routes of the mainstreams of mammalian faunal influx in the Indian subregion

to set aside sufficient representative biotopes to extend protection to a good number of species and their habits. The concept of such resources was also evolved under UNESCO's Man and Biosphere (MAB) Programme.

The foresters in this country, was, however, conscious of the situation and set aside a number of sanctuaries, preservation plots, game reserves and national parks and ensured the protection of flora and fauna. Singalila forests had four preservation plots established in early thirties, besides many others all over Northern Bengal. But such areas were small in size.

Singalila Mountain Range -- It's Speciality

In the eastern part of the Great Himalayan Range, at the border of Nepal, West Bengal and Sikkim, the Singalila Range rises from Kanchanjanga (8335 m) and runs southwards for about 100 km to the northern fringe of the Gangetic plain. The eastern, Sikkim side of the Singalila Range is the valley of the teesta river, the western, east of Nepal side of the valley of the Tamur river, one of the tributaries of the Kosi river. The physiography of the hill ranges on both the eastern and the western sides of Singalila Range are similar to each other. Its situation is unique, where various phytogeographical and zoo-geographical zones meet. It has sub-alpine man-made grasslands that merge higher and higher upto the snow line beyond Singalila peak (Sikkim). Its temperate forests are contiguous with sub-tropical and tropical forests of Tonglu, Ghumsimana, Senchal, Sureil, Chattakpur, Mahaldoram and Mahanadi forests merging to the Sal forests of Chamta and Champasari, Baikunthapur.

The Himalayan Influence

From time immemorial the mighty Himalayan mountains (Singalila Range is a part of it) with their snow-capped peaks, and extremely rich flora and fauna, sweeping panorama of majestic beauty and grandeur, glaciers, rivers, deep gorges have been exerting a tremendous influence on the very existence of Indian people.

The Himalayan ranges which are symbol of celestial splendour is 2400 km long from Kashmir to Arunachal Pradesh and 240 km to 400 km wide with a number of peaks rising well over 7200 M constitute the most striking feature in the Geography of the World. Singalila forests are flanked between Nepal to the west and Sikkim to the east and north virtually at the Trijunction between central and Eastern Himalayas and occupy a Key position both from the view point of Zoogeography and Phytogeography.

Sikkim's Singalila Hills – A Potential Site for National Park

Essentially, national parks are maintained for the preservation of outstanding features : these are dedicated to the primary purpose of preserving, for public enjoyment, preservation of wild life, regulation of stream flow through watershed protection. Outstanding scenic, geologic and other natural wonders are also included. These should be administered to provide for public enjoyment "in such a manner and by such means as will leave them unimpaired for the enjoyment of future generations". National Parks are, therefore, in a sense, great outdoor museums.

Definitions put forward by different countries, however do differ but not substantially. Singalila forests conform the aforesaid qualifications of national park as defined by them.

A nature reserve is set aside for the protection of the aggregate of species contained therein, as well as the supporting physical environment. Putting it another way, the purpose of such reserve is to maintain, hopefully for perpetuity, a highly complex set of ecological, genetic, behavioural, evolutionary and physical processes and the coevolved, compatible populations which participates in these processes. It must be borne in mind that these reserves are not meant to substitute established National Parks and Sanctuaries but may often coinside partly. It must not be confused with Biosphere Reserve where conservation of natural areas and genetic resources are involved. Biosphere reserve a present day concept is to serve for present and future human use the diversity and integrity of biotic communities of plants and animals and to safeguard the genetic diversity of species. However the rationale of Biosphere Reserves is based on the concept of conservation of life in its totality, in situ, so that it should serve as a referral system for monitoring and evaluating changes in natural ecosystems. The Primary objectives of Biosphere Reserve are --

(*i*) to conserve for present and future use the diversity and integrity of biotic commodities of plants and animals within natural ecosystem and to safeguard the genetic diversity of species.

(*ii*) to provide facilities for training and education,

(*iii*) to provide ecological research.

Singalila Range has such basic resources for both such areas. Potentialities of the area are as under :

- The Singalila including part of adjacent Sikkim Himalayas form one of the richest biogeographic region of the Indian sub-continent.
- The region includes some unspoilt natural vegetation.
- This region forms Bio-geographers most interesting and complicated subject in the study of endemism, migration, zonation; and adaptability of species.
- The region represents some unique species of temperate and alpine forests, has a wide variety of biota of various altitudes; has eco-system diversity, naturalness and effectiveness as a conservation unit.
- The region has some rare and endangered species.
- The region is one of the richest avifaunal zone of the world where 550 species (approx.) covering wide range of groups thrive within a small space. A large number of birds are local migrants, others are passage migrants.
- The insect population is extremely rich and greatly diversified. It has vast population of Coleoptera, Diptera, Isoptera, Lepidoptera, Hymenoptera, orders of insects. The fauna in general exhibits affinities with Burma, Thailand, South China and Japan.
- The fish fauna is zoogeographically rich and connects Indo-China, Indo-Malayan and Indian sub-region of oriental region.
- The region is very rich in Lichen, Moss and Liverwort flora.
- The flora primarily exhibits Sino-Japan, Indo-Chinese and Malayan elements. Tibetan, Central Asian and some European elements are also evident at high altitude flora.
- The flora exhibits a high degree of endemism.

- The region has some primitive plants.
- The region is still yielding more Taxa showing that the flora has not been explored completely.
- The region has several species of mammals including some rare ones which show tridimensional zonations. This region lies at the confluence of Ethiopian, Palaearctic, Mediterranean and Indo-Malayan flora and fauna and possess a number of interesting components.
- The region has more than 70 species of snakes and more than 200 species of amphibians.
- The region has many species of Orchids, Ferns and other plants which are rare or endangered.
- The area is connected with a complex mountain chain rising to tremendous heights interpersed with deep transverse river gorges and wide longitudinal valleys.
- The region is extremely rich in medicinal plants and also rich ethnobotanically.
- Owing to various topographical, latitudinal and ecological gradients the region has been the theatre of evolutionary development of several angiosperms.
- This region has wild relatives of cultivated plants.
- This region has immence genetic resources for agri-horticultural research and development.
- Richness of timber plants and profusion of flowering plants of assorted colour are treasure of the region.
- Array of colours of Rhododendron flora which offer a Kaleidoscopic view are special feature of scenic beauty ; so are the dwarf and stunded growth of plants which are features of high altitude plants.
- Epiphytic flora especially the ferns and lichens are natures masterpieces of creation in the locality. Physiological, morphological and biological importance of epiphytes are also the features of the locality.
- Diversity of species in both vertical and lateral planes, variety of habitats and high density of species are also special features.
- The area commands a panoramic view of series of mountain

ranges culminating in world's highest peaks like Mr. Everest, Mt. Lotse, Mt. Makalu and Mt. Kanchanjanga, etc.

- High altitude insect population and their adaptations are also special feature of the locality.
- Alpine pastures with grass -- sedge association, bright colouration of flowers, and a variety of herbaceous flora are objects of study and beauty.
- Other special features of the zone are the juxtaposition of many biotopes or life zones from temperate cold to alpine snow -- contiguity of forests from plains to 12000' (3650 m.) and above -- all telescoped within a straight line distance of 50 km. that has made the tract floristically and faunistically a rare gift of nature.
- The floral diversity parallel the mammalian and other micro and macro -- faunal diversity. Phenophase, Phenodynamic and Phenometry testify to the diversity and richness of the flora.
- Plant-animal interaction (intimate and reciprocal relation) is widely evident in the zone.
- The variety of flora and fauna, the magestic beauty of the mountains and panaromic sight of the snow clad hills are joy for ever and a rare attraction of the tourists.
- The rich varieties of flora and fauna and their inter-relationship provide inexhaustible opportunity for study and research ; the Forests are nature's richest laboratories.

All these qualities enumerated characterise and quantify the area as a special habitat which deserve protection.

Endemism

A good number of species in the mountain are endemic. Notable genera are *Pimula, Rhododendron* besides many species specially the orchids.

High endemism (3200 dicots and 1000 monocots) in Tropical Himalayas show affinities with Indo-Malayan elements. The temperate flora exhibits Sino-Japanese elements. Alpine flora, however, includes Tibetan and Siberian elements. Endemic flora, conditions favourable for survival of primitive plants and speciation on the one hand and ideal habitat for colonization of foreign elements are surely of great significance for conservation of "germ-plasm" and its utilisation in improvement of economic and ornamental plants.

Of the elements of the flora of India, the Malayan is dominant. African, Australian and American elements are less represented than the European flora. In Sikkim - Darjeeling area, 50 per cent of its flora is endemic; others are Japanese 14 per cent, North-American 7 per cent, Australia, Chinese 6 per cent Malayan 4 per cent, Europe 4 per cent, South America 3 per cent, Tropical Asia 3 per cent, Central America 2 per cent, Africa 0.5 per cent and Burma 0.5 per cent. Of the ferns (580 spacies) in India more than 240 are Eastern Himalayan. Of the Mosses, more than 32 per cent are endemic. Of the thallose liverworts 50 per cent are endemic.

Western Himalayas have many floristic elements common with Eastern Himalayan flora (*Ranunculus, Gentiana, Swertia, Anemone* species etc.). Many plants have affinities with Siberian flora (*Potentilla, Pedicularis, Lonicera* etc.). *Juniperus* and *Artimesia* show affinity with central Asian flora. Plants of Papavaraceae and Fumeriaceae show affinity with west Asian flora. In Darjeeling area species of *Schima, Dillenia, Quereus, Bischofia* etc. are Sino-Japanese.

Biogeography

Ecologically and biogeographically the Indian sub-continent is one of the most fascinating region of the world. A great variety of ecological conditions prevail and a great diversity of wildlife exists.

The Singalila region lies at the confluence of Ethiopian, Palaearctic Mediterranean and Indo-Malayan flora and possesses a number of interesting components from each of these realms as well as several peculiarly indigenous genera. The Ethiopian element is represented by the Chinkara, the Hyena and the Ratel, the palaearctic by Lynx, Wolf, Markhor, Urial and Hangal, the Chinese by Red ponds and Musk deer and the Indo-Malayan by the Hoolock Gibbon, the Elephant, the Goat antelope and the Mouse deer.

Wild mammalian fauna of the Himalayas shows tridimensional zonations viz., longitudinal, latitudinal and altitudinal. In each zone, a number of sub-zones can be recognised.

The eastern Himalayan Takin, Tahr, Flying squirrel, Limbless lizards (*Ophisauras*), the Skink (*Leiolopism*), the toadlizard

*(Phrynocephalus)*are all rare animals today. Affinities with Burma, Thailand and South China is predominant.

Abundance of fruits, assorted colourful flowers, profusion of nectars and pollens, myriads of insects, horizontal and vertical contiguity of forest corridor are essential ingredients for sustenance and survival of avifauna and several other animals. Accumulated unburnt leaf on the forest is essential for pheasant, grouse, partridge, fowl and ground dwelling birds. The Singalila forests provide most of these environmental conditions.

Genetic Resources

Immence genetic resources of various agri-horticultural plants occur in the region. The area is rich reservoirs for genetic variability of great agricultural value in a wide group of crop plants. Wild relatives of cultivated plants are *Triticum, Oryza, Cucumis, Cicer, Dolichos, Vigna, Piper, Jasminum, Rose, Camellia, Amomum, Curcuma, Zingiber, Allium,* species etc.

Evolutionary Development in Angiosperm

The region is also rich in timber plants, besides fodder plants, flowering plants and edible fruit bearing plants. There are more than 500 species of tree species in the Terai, Duars and hill forests of Darjeeling district. The project area is rich in temperate tree species especially those of the families *Symplocaceae, Lauraceae, Cupiliferae, Ericaceae* and *Rosaceae.* There are regions of active speciation, hybridization and floral evolution amongst the Himalayan species which contributed to richness of the flora. Owing to various topographical, latitudinal and ecological gradients the Himalayas have been the theatre of evolutionary development of several angiosperms (e.g. *Magnolia, Mlichelia, Talauma, Artimesia, Manglietia, Viburnum, Rhododendron, Morus, Camellia, Buddleia, Saxifrage, Lonicera, Helwingia, Primula, Taraxacum,* Orchids etc.).

Beautiful Landscape of Flowers

The temperate and alpine flora comprises the most beautiful flowering plants (*Primula, Rhododendron, Meconopsis, Magnolia, Viburnum, Hydrangea, Gentiana, Camellia, Cotoneaster, Anemone, Viola, Iris, Spiraea* etc. Ten important plant families of Darjeeling --

Sikkim Himalayas are Orchidaceae, Graminae, Fabaceae, Asteraceae, Cyperaceae, Uriticaceae, Scrophulariaceae, Rosaceae, Rubiaceae and Euphorbiaceae. Most conspicuous are the flowers of *Magnolia campbellii, Flowers of Oaks, Elders, Pyrus, Laurals, Maples* and *Ex-bucklaneia* species are also conspicuous. Best Known Orchids belong to the genera *Cymbidium, Coelogyne, Dendrobium, Eria, Paphiopedium, Pleone, Calanthe, Arundina, Vanda, Phaius, Bulbophyllum, Liparis, Habeneria* and *Cirrhopelatum.*

Medicinal Plants

The region is extremely rich in medicinal plants, some of the genera and species are - *Aconitum, Heterophyllum, Dioscorea* sp., *Digitalis* sp., *Nardostachys jatamansi, Swertia chirata, Podophyllum emodi, Acorus calamus, Rheum* sp. besides very many others.

Some Primitive Plants

There are several primitive genera (phylogenetically) which are *Tetracentron, Alnus, Magnolia, Betula, Ex-bucklandia* etc. Many curiosities like *Drosera* and *Balanophora* occur in the zone. The Himalayas including the Singalila region are still yielding more Taxa showing that the region has not been explored completely.

Quantitative Analysis of Floral and Faunal Resources

As there is abundance of floral and faunal species, it would be a long list to complete at the characteristics of these innumerable species. However, the subject is grouped in two sections (A) and (B), the former relating to flora and the latter relating to fauna.

A. Floral Characteristics

Foliage :

Leaf Surface	–	30%	–	below 800 sq. mm.	Size of leaves
	–	26%	–	from 801 to 1800 sq. mm.	were measured and classified
	–	36%	–	from 1801 to 2500 sq. mm.	into four classes
	–	8%	–	about 2500 sq mm.	

Types of leaves	– 51.5%	– Simple alternate	Types of leaves were grouped into four classes
	– 27.9%	– Compound alternate	
	– 13.0%	– Simple opposite	
	– 7.6%	– Compound opposite	

Climbers, Eniphytes (per heactare) in Wet Mixed Forests.

500 Nos.	– Climbers
100 "	– Loranthus
285 "	– Epiphytes
100 "	– Hard fungi; except for gregarious Daedalia, Polystictus, Fomes and some other species on slash

In temperate forests one sq. km. has about 100 species of epiphytes. This has special significance to avifauna population.

Diversity and Density

(*a*) *Wet-mixed Forest*

Morghat Reserve	250 trees per hectare	This dense crop of tree vegetation in the top tree storeys mingled with dense herb and shrub layers form an ideal habitat condition for wide species of Wildlife population
Lalaguri Reserve	213 trees per hectare	
Damanpur Reserve	225 trees per hectare	
Rajabhatkhowa Reserve	275 trees per hectare	
Apalchand Reserve	190 trees per hectare	
Udalbari Reserve	200 trees per hectare	

(*b*) *Canopy Layer and Percentage*

Canopy layer	No. of species.	Percentage	
Top storey	34	25.8	This importance of middle storey to various bird, squirrel, reptiles and lizards and other animal population is significant.
Middle storey	64	48.4	
Lower storey	34	25.8	
	132 sp.	100%	

(c) *Vegetation Survey. (In Darjeeling Himalayas over 6000 sq km)*

In Dicot :

Total Tree species	—	540 species
Herb species	—	743 species
Shrub species	--	321 species
Climbing shrub	--	105 species
Large climbers	--	111 species

In Monocot :

Shrubs & Trees	--	307 species
Herbs	--	508 species.

50 per cent of entire Dicot and Monocot is herb ; 23 per cent of the herbs (1251 species) is in the plains.

There are 205 species of fern and fernallies, 11 species of Gymmosperms, over 300 species of Moss and Lichens.

Assemblage of these innumerable species in such a small area is immediately significant to wildlife.

(*d*) *Scattered Shorea robusta Forests*

Damanpur	—	325 trees/hectare.	These are basically wet mixed forests and the selected spots mentioned on the left form one of the best zones for Wildlife population
Jainty	—	312 trres/hectare	
Bengdubi	—	312 trees/hectare	

(*e*) *Pure Shorea robusta Forests*

—	295 trees/hectare	This area do not have enough number of middle and lower storey miscellaneous species and therefore not so much significant to Wildlife as other areas.

(*f*) *Fruiting and Flowering Season vis-a-vis Wildlife*

Flowering season of Monocot families in general are spread out from April to August with peak period in June.

Among the dicot families the following are variations :

Polypetalae — Peak period of flowering is April and May, while there is a smaller spurt in September-October.

Gamopetalae – In Rubiaceae peak period is June-July with a small spurt in April.
– In Asteraceae peak period is July-August with a small spurt from May to October.
– In Lamiaceae peak period is September-October.
– In Scrophulariaceae peak period is October, a small spurt is from June to October.
– In Verbenaceae peak period is from May to September.

(g) *Sequence of Flowering in the Hills*

This region is exceedingly rich in conspicuous flowering species of herbs, shrubs, besides trees. A visitor will find the following species in bloom during April to October :

In April – Chestnuts, *Hydrangea, Magnolia* (they throw out leaves and flowers). Purple *Stauntonia,* white and pink *Daphne* cannabina adorn forest floor. *Arisaema* appear in profusion along with *Paris* and *Polygonum.* Yellow strawberry, *Dispora, Melissa* and *Isopyrum* are common. White *Coelogyne* adorn moist rocks; *Cymbidium* grows on trees. *Rhododendron* species start flowering.

In May – Flowering of Taxus, Mapple, *Photonia* is followed by *Symplocos, Alder,Saurauia* and *Engelhardtia.* Flowering of *Vaccinium, Indigofera, Cotoneaster, Ribes* and *Gaultheria* shrubs start. Conspicuous herbaceous species that Flowers are *Viola, Anemone, Potentilla, Fagopyrum, Primula, Gentiana* and *Androsace.*

In June – Common flowering shrubs are *Polygala, Spiraea, Jasminum, Neilla, Leycesteria* and *Polygonum.* The herbs that blossom are *Gerania, Parochaetus, Potentila* etc. Berries of *Rubus* and *Elaeagnus* ripens. Purple cones of *Abies,* red-brown cones of *Larix,* barries of *Junipers* are conspicuous. Primulas having white, yellow and purple colour are abundant. Grasses and sedges start flowering.

In July – Shrubby species of *Hypericum, Deutzia, Balsams, Begonia, Torenia, Codonopsis, Thalictrum, Orchids* (terristrial species of *Calanthe* and *Habenaria), Valeriana, Notochacte, Convolvuulus* (purple flower) and *Polygonum* (white) flower. Sparingly occurs are species of Balanophora.

In August -- Various species of ferns throw their foliage, yellow flowered *Trichosanthes, Anemone, Hedychium* commonly occur.

In September-- Abundance of Labiatae (*Plectranthus, Elsholzia, Colquhounia, Brunella, Melissa* species). Some *Anaphalis, Osbeckia, Polygonum* and many Urticaceous species dominate.

In October -- Common flowering plants are *Prinsepia, Crow fordia, Clematis, Camellia, Eurya, Symplocos, Prunus,* Fragrant *Olea* and *Ex-bucklandia.*

Several grasses flower in October-November.

(*h*) *Fruiting*

Analysis of fruiting seasons however shows that fruiting of species occur throughout the year (analysis of 57 important families). The fruits are of different sizes and colours : they are pulpy, arillous, sweet and bitter.

Wide diversity of plant species has created a large store house of flowers, nectar and pollen and also of fruits which sustain insects, birds and animals beside meeting some need of human population.

(*i*) *Ground Vegetation Survey in Plains*

Analysis of 0.4 hectares of forests show the following (number of species) :

Types of species	Sal forests	Riverain forests	Dry mixed forests	Plantation
Grasses	9	8	6	5
Ferns	5	2	1	2
Fernallies	-	1	1	1
Sedges	1	2	2	1
Monocot	10	4	8	4
Climbers & Climbing shrub	23	21	11	8
Herbs & shrubs	17	19	12	9
Tree species regeneration.	47	15	31	25
Total species	112	72	72	55

Most of the species in Sal forests are of high frequency. In Plantation areas (monoculture) species diversity is less.

(*j*) *Ground Vegetation in the Hills*

(a) Predominant and common families and species in the ground vegetation are as follows (number of species in bracket) :

Urticaceae (12), Fabaceae (11), Asteraceae (11), Filicales (11), Graminae (8), Lamiaceae (6), Rosaceae (5), Scitaminae (4), Gentianaceae (5), Acanthaceae (4), Umbelliferae (4), Liliaceae (3), Commelinaceae (3), Orchidaceae (3), Begoniaceae (3), Cyperaceae (5).

(b) An analysis of ground vegetation shows as follows :

1 × 1 sq.metre quadrat-8-18 species ; 180-350 individuals ; *Selagenella* forming a mat on ground floor.

2 × 2 or 4 sq.mt. quadrat - 24 species, 3 × 3 or 9 sq.mts. quadrat - 36 species, 4 × 4 or 16 sq.mt. quadrat - 46 species.

It is not possible to count individuals beyond the size 4 sq.mt. Vital zone is from 1000 to 1500 metres where various species are aggregated.

This diverse habitat will bear a direct linear relationship with bird species diversity in the wet mixed forests of the study area.

B. Avifaunal characteristics

Salim Ali's paper on Himalayan Ornithology (1981) asserts the richness of avifauna of Sikkim and Nepal borderland. B.H. Hodgson (1800-1894), he considers, as the 'Father of Himalayan Ornithology'. Salim observed close association between the genus *Garrulax* (Laughing thrush) with the shrub *Rubus* (raspberry); Also the occurrence of such association in south Indian hills. In this connection Dr. Hora's Satpura Hypothesis may be mentioned. The great Pied Hornbill (*Buceros bicornis*), Frogmouth (*Batra chostomus*, Fairy Blue Bird (*Irena sp.*), Lizard Hawks or Bazas (*Aviceda* sp.) and Rufous billed Hawk Eagle (*Hieraaetus Kienerii*) all these species which occur in the Himalayas also occur in South India.

Salim Ali records 2100 avifaunal species in the Indian Sub-continent, 300 species of which he considers are winter visitors from the Palaectic region north of the Himalayan barrier; some of the species are red start, ducks, cranes, wading birds, geese, fly catchers, warblers, thrushes, swallows, robins, wagtails, pipits, finches etc.

The Sikkim and adjoining areas are one of the richest avifaunal zone of the World. About 550 species of avifauna have been recorded in the area. Of these, more than 50 per cent are passerine birds, the largest being the Timaliidae (Laughing thrushes, Babblars etc.) with 61 sp. and Turdidae (Chats, Thrushes etc.) with 56 sp., the Sylviidae (Warblers) with 60 species and Musicicapidae (Flycatchers) with 27 sp. and the Fringllidae (Finches) with 22 sp. Some of the birds are exceptionally beautiful. Woodpeckers (15 sp.), Cuckoos (16 sp.) and Owls (14 sp.) and Hornbills are special features of the zone. Accipitrine birds number 40 and include five Himalayan Lammergeyers and Hodgsons feathertoed hawk eagle. There are about 12 species of pigeons and dove.

The game birds are of 8 species and consist of the Indian red jungle fowl, the black-backed pigeons, the monal, the blood pheasant etc. There are several species of partridges and quails.

A large number of birds are local migrants moving up and down the hills according to season. Others are passage migrants using the area for movement to the north in Siberia, Tibet, China to the plains of India. Thrushes, Flycatchers, Warblers, Shrikes, Swallows, Cuckoos, Pigeons, Woodcock etc. all birds of higher mountains that descend to the plains in autumn.

Great numbers of migrating Finches Larks, Pipits, Wagtails, Swifts, Redstarts, a few species of Eagle, Harrier Shipa, Quail and Duck stop for a short time in the district. Huge flocks of Goose and Crane fly over the district. Typical birds of Temperate hills 1550 m -- 2750 m are Yellow naped luxlus, Striated bulbul, Dark grey bush chat, Chestnut bellied rock thrush, Red tailed riala, Red headed tit, Red headed laughing thrush. Typical birds of sub-alpine hills 2750 m to 4225 m are Crimson horned pheasant, Laughing thrush, Nut-craker, Crested black Tit, The common hill partridge, Yellow backed sunbird, Snow partridge, The Wood cock, The Ashy wood pegion, The Great parrot bill and The Fulvous fronted parrot bill.

Some of the high altitude birds are Finches, Parrot bills, Chats, Pheasants, Pigeons, Sunbirds, Tits, Warblers, Cuckoos and Flower peckers.

Some birds with long latitudinal ranges are :

The Indian Sparrow hawk, The Himalayan griffon vulture, The Lammergeier or bearded vulture, The Rufous Turtle dove, The Large hawk cuckoo, The Himalayan cuckoo.

There are abundance of flowers of various colours, sizes, shapes, various types of pollens and nectars, various sizes and colouration of fruits, blossoming of flowers and ripening of fruits round the year, various sizes of insects in the zone which has created an ideal condition for survival of various fungivorous and insectivorous birds.

Migration of Avifauna

The migration of ducks, geese, cranes in India occurs through Siberia-China over Tibet into Eastern India. On this long march the vast migratory bird population need transit facilities in the Himalayas. The conservation of habitat in the Himalayas, as such, is essential.

Avifauna of Darjeeling and Adjacent Areas

(*a*) *Size relationship between various fungivorous avifauna and seeds and fruits*

Sizes of fruits available in the forests were measured and classified into four groups and percentage of each class was calculated as follows :

(*i*) Below 0.5 cm - 60.0%, (ii) between 0.5 and 1.0 cm -- 28.0%, (iii) between 1.0 and 2.0 cm -- 8.0%, (iv) Over 2.00 cm -- 4.0%.

Fungivorous birds (Pigeons, Doves, Pheasants, Quails, Horn bills, Broad bills, Barbets, Mynas, Nuthatches, Bulbuls, Finches and Sparrows) were classified into the following sizes and percentage of each class was calculated as follows :

Bird length size							
	7.5 cm	to 15.0 cm	-	30.0%,	16.0 cm to 25.0 cm	-	35.0%,
	26.0 cm	to 30.0 cm	-	8.0%,	31.0 cm to 35.0 cm	-	7.0%,
	36.0 cm	to 45.0 cm	-	6.0%,	46.0 cm to 55.0 cm	-	6.0%,
	56.0 cm	to 65.0 cm	-	2.0%,	66.0 cm and over	-	4.0%.

While the smallest fruit size constitutes 60 per cent of total, the smallest sizes of fungivorous avifauna form 65 per cent of total ; this is had the nature strikes a balance. But the size, colour and acceptability of insect by avifauna is a complex phenomenon and need a thorough study.

(b) Colouration of Flower vis-a-vis Avifauna

An analysis of colouration of flowers show the following classification.

White flower -- 28.5% ; Yellow flower -- 24.0% ; Pink flower -- 10.0%, Red flower -- 12.0% ; Others -- 25.5%.

Colouration of birds in the same zone show the following percentage :

Brown -- 43.0% ; Green -- 6.0% ; Ashey grey -- 12.0% ; Pied (Black & white) -- 9.0%, Black -- 8.0% ; Mixed bright -- 5.0% ; Red -- 3.5%; Yellow -- 4.0% ; Blue -- 3.5% and Others -- 6.0%.

Flowers colouration, time of bloom and scent have a positive correlation with the visit of avifauna for collection of nectar and pollen. It has been found that green coloured flowers are generally a tomophilous, red and orange coloured and odourless flowers are generally bird pollinated, Dull coloured and sourish scented flowers are bat pollinated. Night blooming flowers are generally white and strongly scented. The moths avoid light coloured flowers. Diptera prefers flowers having smell of fruits. Hymenoptera likes over ripe fruits. Such studies have wide scope in the forests of Darjeeling, Sikkim and adjoining areas.

(c) Size Classification of Avifauna in General

In para (a) the sizes of fungivorous birds upto 25 cm. have been shown as 65 per cent. In a fresh classification of avifauna (Omnivorous, insectivorous and fungivorous) in general the following size distribution has been recorded :

Over	7.5 cm	- 10 cm	- 12.4%;	over 10 cm	- 15 cm	- 24.4%;
over	15 cm	- 20 cm	- 12.8%;	over 20 cm	- 25 cm	- 14.0% ;
over	25 cm	- 30 cm	- 6.8%;	over 30 cm	- 35 cm	- 0.8% ;
over	35 cm	- 40 cm	- 2.8%;	over 45 cm	- 55 cm	- 5.2% ;
over	55 cm	- 65 cm	- 5.2%;	over 65 cm	- 75 cm	- 3.7% ;
over	75 cm	- 1.9%.				

SIZE RELATIONSHIP BETWEEN FRUGIVOROUS BIRDS TO SEEDS / FRUITS

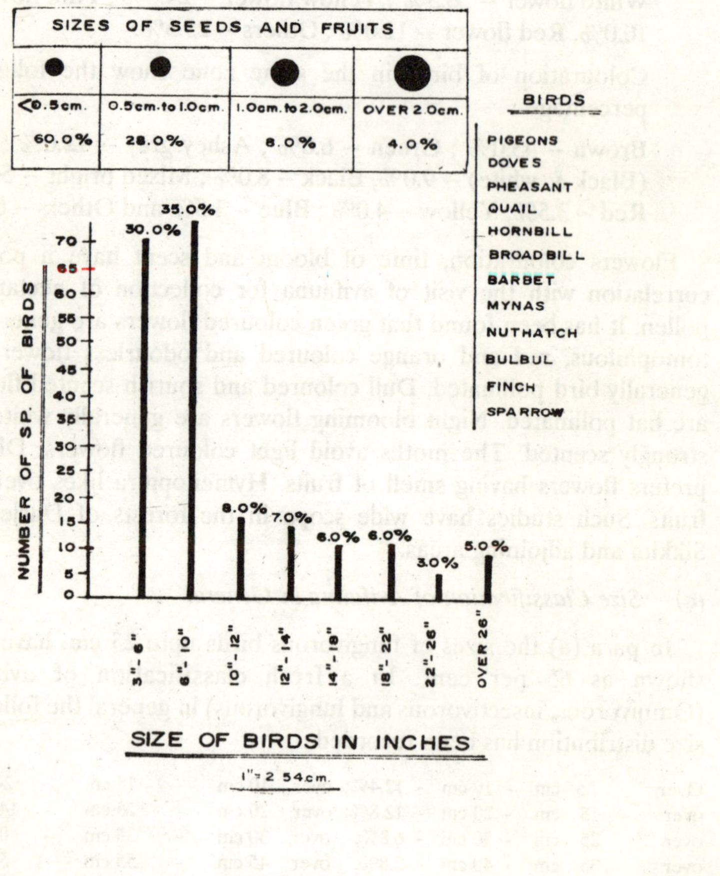

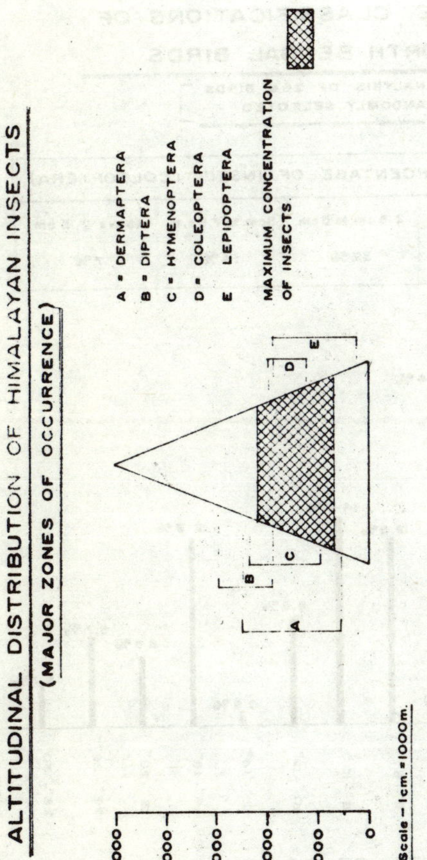
ALTITUDINAL DISTRIBUTION OF HIMALAYAN INSECTS
(MAJOR ZONES OF OCCURRENCE)
A = DERMAPTERA
B = DIPTERA
C = HYMENOPTERA
D = COLEOPTERA
E LEPIDOPTERA
MAXIMUM CONCENTRATION OF INSECTS
5000
4000
3000
2000
1000
0
A
B
C
D
E
Scale – 1cm. = 1000m.

SIZE CLASSIFICATIONS OF NORTH BENGAL BIRDS

[ANALYSIS OF 266 BIRDS RANDOMLY SELECTED]

SIZE PERCENTAGE OF INSECT (COLEOPTERA)			
<2.5 cm.	2.5 cm. to 5 cm.	5cm. to 7.5 c.m.	Above 7.5 cm.
47%	32%	14%	7%

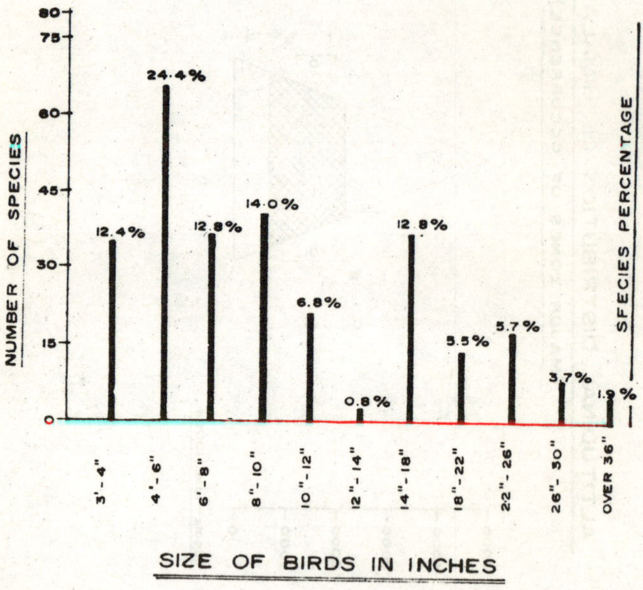

1" = 2.54 cm

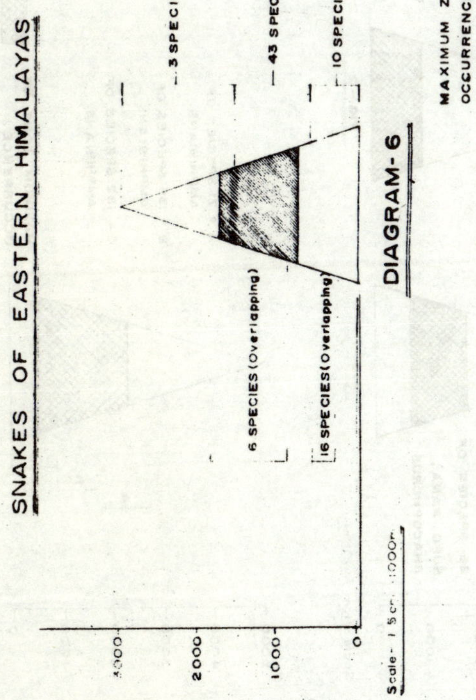
SNAKES OF EASTERN HIMALAYAS
3 SPECIES
43 SPECIES
10 SPECIES
6 SPECIES (Overlapping)
16 SPECIES (Overlapping)
2000
1000
0
MAXIMUM ZONE OF OCCURRENCE.

DIAGRAM- 6

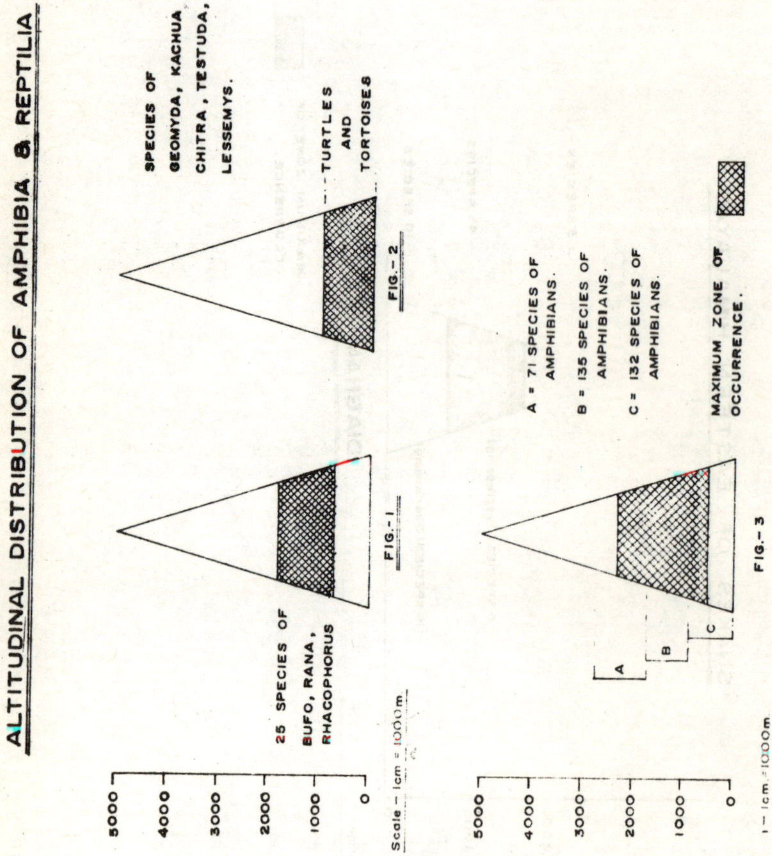
ALTITUDINAL DISTRIBUTION OF AMPHIBIA & REPTILIA
5000
4000
3000
2000
1000
0
25 SPECIES OF
BUFO, RANA,
RHACOPHORUS
Scale – 1cm = 1000m
FIG.- 1
SPECIES OF
GEOMYDA, KACHUA
CHITRA, TESTUDA,
LESSEMYS.
TURTLES
AND
TORTOISES
FIG.- 2
5000
4000
3000
2000
1000
0
A
B
C
1 – 1cm = 1000m
A = 71 SPECIES OF
AMPHIBIANS.
B = 135 SPECIES OF
AMPHIBIANS.
C = 132 SPECIES OF
AMPHIBIANS.
MAXIMUM ZONE OF
OCCURRENCE.
FIG.- 3

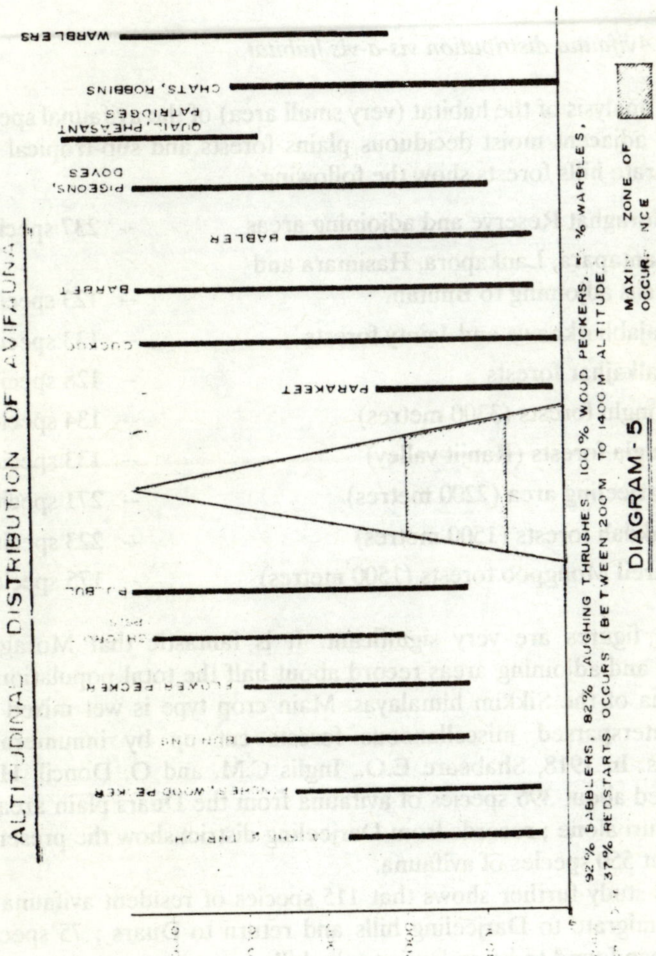
ALTITUDINAL DISTRIBUTION OF AVIFAUNA
DIAGRAM 5
WARBLERS
CHATS, ROBBINS
QUAIL, PHEASANT
PATRIDGES
PIGEONS,
DOVES
BABLER
BARBET
CUCKOO
PARAKEET
BULBUL
FLOWER PECKER
31% REDSTARTS OCCURS BETWEEN 200M TO 1400M ALTITUDE

This classification also shows that 63.6 per cent of the avifauna are short sized and they need smaller fruits and smaller insects for their food. It seems nature has kept provision of edibles proportionate to size.

(d) Avifauna distribution vis-a-vis habitat

An analysis of the habitat (very small area) of the avifaunal species in the adjacent moist deciduous plains forests and sub-tropical and temperate hills forests show the following :

a. Moraghat Reserve and adjoining areas -- 237 species
b. Hantapara, Lankapora, Hasimara and areas adjoining to Bhutan. -- 123 species
c. Rajabhatkhawa and Jainty forests -- 133 species
d. Dalkajhar forests -- 128 species
e. Tonglu forests (3300 metres) -- 134 species
f. Singla forests (Ranjit valley) -- 133 species
g. Darjeeling area (2200 metres) -- 271 species
h. Takdah forests (1500 metres) -- 223 species
i. Sureil Mongpoo forests (1500 metres) -- 175 species

The figures are very significant. It is fantastic that Moraghat forests and adjoining areas record about half the total population of avifauna of the Sikkim himalayas. Main crop type is wet mixed sal and intersparsed miscellaneous forests cut up by innumerable streams. In 1918, Shabeare E.O., Inglis C.M. and O. Doncil, H.V. recorded about 396 species of avifauna from the Duars plain area of Jalpaiguri alone ; records from Darjeeling district show the presence of about 550 species of avifauna.

This study further shows that 115 species of resident avifauna of Duars migrate to Darjeeling hills and return to Duars ; 75 species have been found to be endemic to the hills.

If this figure is compared to the areas rich in avifauna, we find that the Duars and Darjeeling areas are as much rich (survey figure in brackets) as those of Costa Rica (269 sp. in 7 sq. miles), Panama (175 species in 6 sq.miles), East Congo (128 sp. in 20 sq. miles), South Nigeria (117 sp. in 25 sq. miles) and Malayasia (175 sp. in 4 sq.

miles).

Altitudinal distribution of Nuthatch, Finches, Wood pecker, Laughing thrush, Parrot bill, Chats and Robbins, Pheasants, Partridges, Awaits, Pigeons and Doves, Tits, Babblers, Warblers Parakeets, Cuckoos, Barbets, Bulbuls, Chloropsis, Flower pickers, Sunbirds and Bantings show that the hills areas between 600 metres to 1500 metres is the most vital zone for being habitat for these birds where they record their maximum concentration. To record specifically, the lower half of this zone is more vital. The diversity of flora is also maximum in this zone.

(e) Aquatic plants and avifauna

A rough analysis of most of the swamps shows that there are at least 100 species of vascular aquatic plants. Besides these vascular plants, there are innumerable species of algae flora (phytoplankton) and fauna (zooplankton). Dispersion of such vascular flora are Floating -- 20 species, Submerged - 50 species and Marginal -- 30 species.

Besides local waterbirds such as Purple moorhen, Bronze winged jacana, Pheasant tailed jacana, Ibis bill, Sand piper, Little Stint, Wood cock, Snipe, Tern, Cormorant, Stork, Heron, Egret, Bittern etc. there are plenty of migratory water birds.

About the migratory duck population, an assessment is possible from various records (percentage of birds shot) on Shikar sports which is as follows :

Common Teal	-- 55.0%	Redeyed pochard	} 2.0%
Garganey teal	-- 15.0%	Ruddy shell drake	
Shoveller	-- 10.0%	Spot bill	
White eye	-- 6.0%	Baikal teal	
Gadwall	-- 6.0%	Comb duck	
Wigeon	-- 3.0%	others	
Pintail	-- 2.0%		
Mallard	-- 1.0%		

But these are from past records. Present record will obviously show decline in abundance and dispersion in the number of species.

Fish Fauna

The fish fauna of Eastern Himalayas is very rich compared to that of rest of India. This area is zoogeographically important because it connects the Indo-Chinese, Indo-Malayam and Indian sub-region of large Oriental region. Shebbeare and Shaw have identified 136 species of fish in the plains and hills of Darjeeling. Late Hore and Gupta reported 58 species of fish from Kalimpong and Siliguri forests areas. Most important game fishes are *Barbus dukai*, *B. putitora, B. tor, Bagarius bagaries* and *Barilius bola*. These fishes have various forms. Barbus has red and copper colour varieties. *Barilius* feed on wide variety of insects, *Barbus dukai* prefers fruits and algae generally, but other species of *Barbus* feed on various food from the bottom of rivers and streams. Over-exploitation, situation in rivers, dynamiting and use of fish-poison in water have reduced fish population to base minimum.

Leech Fauna

There are six species of leeches belonging to genera *Dibobdella, Haemadispa* and *Hirudinaria.* They grow in profusion in countless numbers and attack all species of domestic and wild animals in the hills. *Haemadipsa montana* has been recorded from high altitude (3000 m), but it is rare.

Amphibian Fauna

Reptile and amphibean fauna in eastern Himalayas have distribution in different altitudinal zones as follows :

between	0	--	900 m	--	132	species
"	900	--	1800 m	--	13	"
"	1800	--	2700 m	--	71	"
"	2700	--	3700 m	--	38	"
	and above					

So far cataloguing and taxonomic descriptions of amphibians have been made and written, but no detailed ecological work has been done. The wealth of insect and amphibian life and their marvellous adjustment with the special environmental condition of high altitude should be of the special interest.

Tylotricton verrucosus (Salamandar) which is a rare species occurs about 2600 m. *Varnus salavator* occurs from plains upto 2300 m.

Platuyurus sp. extend from 900 -- 2800 m., *Leiolopisma Sikkimensis* occurs from 900--3100 m., *Lagossoma maculatum* occurs from 2300-3100 m., and *Calotes versicolor* climbs upto 2600 m.

Of the frogs some species of *Rana, Philautus, Rhacophoras* and *Bufo* occur in upper temperate and alpine Himalayas. Typical tree frog of upper temperate hill is *Philautus annandalii. Bufo melanostictus* occurs upto 3500 m. *Rana* sp. (*R. monticola, R. formosa, R. himalayana, R. liebigii, R. limonocharis)* occur from plains to 2300 m. *Geomyda, Kachua, Testudo, Lissemys, Trionyx* and *Chitra* occur in the foot hill forests of the district.

Snake Fauna

Darjeeling hills and plains have at least 76 species of snakes. Although maximum concentration of various species centres round 800--1570 m., several species predominate in the planes and foot hills, others inhabit 1570--2200m ; few species also extend in the upper temperate hills and alpine pastures. Middle hills have wall's blind snake, Stoticzka's blind snake, Jerdons blind snake, Burmese Diards blind snake, Collared dwarf snake, Anderson's keelback, The buff striped keelback, Goggle eyed keelback or hooded tree snake, Black bellied worm eating snake, The black-bordered rat snake, The ringtailed Dhaman, Common wolf snake, King Cobra (upto 2300 m.), The Large spotted viper, The Himalayan pit viper species which are worth mentioning.

Some species of the genera *Mabuya Lygosoma, Elaphe, Callophis, Nija, Vipera, Trime, Ptyas, Zoaays, Liopelies* occur in upper temperate and sub-alpine hills.

Mammalian Fauna

The Singalila range and adjacent areas have 80-90 species of mammals. Formerly, tigers and leopards were sometimes sighted at an altitude of 3200 m. Several species of cats such as the Leopard cat, the Golden cat, the Fishing cat, the Himalayan jungle cats and five species of Civet cats are found in the area.

Indian sloth and Himalayan bear are quite common. Coloured leopard, snow leopard and Pandas have been observed at about 2600 m. in Singalila Range, but barring the Pandas other two species are rare.

Weasels, Marten, Squirrels, Bats are common in the area ; the Himalayan flying squirrels and Assam giant squirrels are beautiful animals of the area. Bisons do occur in the foot hills but do not climb upper hills. Elephants too do not climb in the upper hills.

Bengal barking deer can be found at 2600 m. but not the other species of deer. Serow and Goral do not climb more than 1600 m. altitude. The Indian Wild boar sometimes climbs upto 2600 m.

Plant-Animal Interaction

Colouration, shape, size and scent of flowers, scent and size of fruits are correlated to various evolutionary processes and behavioural pattern of insects and avifauna species. Bees are sensitive to red, yellow and blue colours ; hive bees prefer blue and yellow flowers ; Bats prefer drab green colours ; Beetles and moths prefer red, purple, white and pink flowers ; Flies prefer scarlet or yellow or red flowers. This zone has all these assorted coloured flowers. All these flowers, their structure, colour etc. are closely criped with various species of insect which interact among themselves.

An analysis of colouration of flowers among tree herb and shrub communities is as under :

Colour	Tree	Shrub	Herb
White	33	33	24
Yellow	23	24	22
Purple	4	18	18
Red	7	10	8
Pink	5	9	13
Blue	8	6	15
	100	100	100

These areas have a very rich mixture of herbs and shrubs ; it is a zone where the herb population goes on increasing, the shrub goes down and the incidence of climber is maximum. Out of 111 species of climbers recorded from the districts of Jalpaiguri and Darjeeling, about 80 species occur in this zone.

Insect

The insect life of the Himalayan forests is extremely rich and greatly diversified. The wooded slopes of higher mountains are populated by a greater abundance of mostly all sorts -- May flies, Stone flies, Grass hoppers, Crickets, Dragonflies, Caddies flies, Bugs, Moths, Midgets, Ghats, Mosquitoes, Fleas etc. The greatest bulk of the species are ground living, wingless and are carnivorous and debris feeder. They live in rock crevices, on moss and on other dwarf plants, on snow edge, in torrential streams of water, ponds and in spring.

From recorded distribution of various species of Coleoptera, Diptera, Hymenoptera, Orthoptera, Lepidoptera, Rhynchota etc. vertical zonation map was prepared showing maximum concentration of insects within short latitudinal range.

The area is exceedingly rich in Lepidoptera. Some of the rare and most beautiful moths occur in this area. Brilliantly coloured beetles also occur in the area.

High altitude insect population and ecology has been studied in Western Himalayas by the Indian Zoologists and information recorded in proceedings of Zoological Survey of India in which bring out very interesting behaviour pattern and adaptation of high altitude insects. Similar study in the eastern Himalayas will reveal various unknown facts on high altitude insects where high light intensity, reduced atmospheric pressure, physical and topographical peculiarities, Oxygen deficiency, ultraviolet radiation, high wind velocity, low temperature, atmospheric humidity and aridity prevail.

Abundance of Coleoptera has been recorded between 1200 m and 1900 m, Dermaptera between 1500 m and 2200 m, Diptera between 1500 m and 2000 m and Hymenoptera between 1300 m and 2000 m, though the distribution of the species attain a much wider range.

Insectivorous Birds Vs. Insect Colouration

In this locality the abundance of moth (surpassing butterfly sp.) goes parallel with abundance of insectivorous birds. White coloured moth is widely eaten by birds, light yellow orange coloured moths are less accepted, but brilliantly coloured ones are not accepted. Moth concentration is maximum in foot and middle hills though quite a number of species do occur in upper temperate zone. Incidentally, the distribution of Moth (Sphingidae) show as follows :

W. Himalayas 62 sp., Eastern Himalayas 135 sp., South India 75 sp., Ceylon 58 sp., Burma 49 sp., Andamans 22 sp. This shows the abundance of moths in eastern Himalayas.

Lepidoptera and Forests Habitat

The butterflies and moths are most conspicuous of all insects in the forests and they not only form exquisite landscape, they are also assets to our forests. About 60 per cent of the total butterflies and moths of India are represented in the North Eastern India. Distribution of Sphingidae (Moth) shows that Burma, Andaman, Ceylon, South India, West Himalayas and East Himalayas have 401 species of which the vast named area has 135 species.

They grow to various sizes, but 70 per cent of the species falls between 5 cm. to 8 cm. largest ones (*Actias*) have size above 17 cm. Colouration of insects and insectivorous birds have direct bearing on the predator prey animal relationship. But further work in the tropical forests is necessary. Colouration of moth shows preponderance of brown colour of various grades and mixtures.

Moth and host plant relationship

Moths and butterflies have specific host plant for their larvae and cocoon formation. The following plant families are of prime importance - *Rubiaceae, Ampelidae, Aroidae, Urticaceae* and *Leguminoceae.* The family Urticaceae is of primary importance to the butterflies of Darjeeling hills. It has been found that as many as 61 species of moths are host on a single plant family which shows importance of some species usually considered a weed by the foresters.

Host Plants of Hawk Moth

An analysis shows the following :

45 species individually are host to 1 family and 1 species.
8 species individually are host to 1 family and 2 species.
3 species individually are host to 1 family and 3 species.
4 species individually are host to 1 family and 4 species.

Butterflies and Host Plant

The Dinaids Larvae feed on Urticaceae, Apocynaceae and Asclepiadaceae. They are common between 900 metres to 1200 metres. Troides and Tros are confined to Aristochiaceae, Papilio on Rutaceae, Pathysa and Zetides on Anonaceae and Lauraceae, *Zetides and Chilasa* on *Machilus odoratissima, Papilio paris* on *Evodia roxburghiana, P. pravidarun* on *Glycosmis orborea, P. liomedon* on *Acronychia laurifolia, P. machaon* on *Umbelliferae*. Many such examples may be cited. This shows that some plants and plant families are specific to particular genera and species of butterflies. Eradication of the families or species from the area will obviously eradicate the species of butterflies altogether.

The following plants and plant families are important hosts of North Bengal butterflies :

Myrsinae, Passiflorae, Urticaerae, Asteraceae, Rubiaceae, Acanthaceae, Moraceae, Fabaceae, Asclepiadaceae and Apocynaceae.

Colouration

It seems the brown or whitish brown colour predominates over 50 per cent of the species ; the balance are colourful, of which some are brilliantly coloured.

A few butterflies move long distance (several kilometres) over regular routes through the forests to collect aminoacid from pollens from particular species of plants to support their egg production. As such no plant species should be termed as weed and outright rejected.

Butterflies are abundant where plenty of larval food in the form of young green leaves are available. Out of 1443 species of butterflies of India, 835 species occur in East India alone. In the study area there are more than 3000 species of plants and obviously adequate food materials.

Coleoptera

Among the insects Coleoptera population seems to surpass all other species in the hills. Recent collections of Z.S.I. between 1000 and 2000 metres about 6572 individuals of Coleoptera against 1321 of other orders (Hemiptera, Hymenoptera, Diptera, Orthoptera, Lepidoptera, Isoptera, Dermaptera, Thysanoptera, Odonata etc.)

The author made size class distribution (expressed in percentage) of Coleoptera of Darjeeling Himalayas and found the following :-

Less than 2.5 cm - 47% ; 2.5 to 5.0 cm. - 32% ;
5.0 to 7.5 cm - 14% ; above 7.5 cm - 7%.

This size class would be significant in the analysis of food habit of insectivorous birds of the area. Distributional study of 270 species of Coleoptera shows maximum concentration between 1250 metres and 1800 metres in the Darjeeling and Sikkim himalayas.

Dermaptera

Maximum concentration of hills species in between 1250 metres and 2200 metres.

Diptera and Ungulata Relationship

The Dipteras are very important in the ungulata (deer, pig etc.) ecology. The species of ungulates are abundant in the plains and foot hills of the Himalayas but detailed figures of analysis are not available.

The altitude between 1500 and 2200 metres is rich in Diptera insects, but they can hardly be correlated in any way with the ungulates in eastern himalayas were they have never been so much in abundance. But this abundance is significant to the Avifauna population and forests as the Dipteras are very good pollinators and are attracted by odours of wide variety of flowers and fruits which are abundant in the forests. The Tabanids need special attention as the animals are sometimes driven out of forests by their massive number attacking them.

Hymenoptera

Analysis of distribution range of about 362 species in the Himalayas show that 1300 to 2200 metres from the vital zone of the insect group.

Some Endangered Plant Species

The following are some of the many other endangered plant species of the area :

Orchids : *Dendrobium densiflorum, Vanda pumila, Paphiopedium insigne, P. ceenustrum, Arundina graminifolia, Cymbidium macrorhizon, C. marcranthon, Arachnanthe cathcartii, C. clarkii, Aphyllorchis montana*.

Ferns and fern allies : *Cyathea gigantea, Osmanda regalis, Alsophila species, Botrychium virginanum, Angiopteris evecta.*

Other Angiosperms. : *Balanophora dioica, Dioschorea deltoidea, Helwingia himalaica, Nardostachys grandiflora, Populus gamblei, P. glauca, Rhododendron dalhousiae, R. edgeworthii, Rush bookeri, Tetracentron sinense* var *himalense, Salix daltoniana, Viscum caudatum, V. album* and *Viscum articulatum*,

b. Very conspicuous ephiphyte ferns of Rhododendron Conifer forests are : *Oleandra wallichii, Polypodium argutum, P. lachnopus, Lepisorus excavatus, L. kashyapie, L. longifolius, Arthromeris himalayensis, A. lechmanni, A. wallichiana Vittaria himalayensis, Chenopteris subfalcata, Phymatodes edenipes, P. oxtloba, Araiostegia, dareiformis* grow on diverse niche viz., on rocks, tree trunks, tree bole, tree top and on other situations. There are several other species which occur in the area.

c. Of the fern allies, conspicuous species are *Lycopodium selago* and *L. serratum.*

High Level Vascular Epiphytes

High level vascular epiphytes form special sinusia of the temperate forests. The base of leaves of such epiphytes form special niche for other species of plants and wildlife to survive. High level bascular epiphytes are also another remarkable feature of the temperate forests. The plants grow either on stones or branches or on the bole of principal tree. Common species are :

Agapetes serpens, Aeschynanthus sikkimensis, Gonatanthus humilis, Cautleya spicata, Globba sp., *Lysionotus serrata, Loxostigna griffithii, Pentapanax leschenauthii, Raphidophora glauca.*

On branches grow *Hynenodictyon flaccidum, Hymenopogon parasiticus, Agapetes saligna, Rhododendron edgeworthii, R. vacconioides, Fagraea ovata, Acer hookeri* and *Rhododendron dalhousie.*

Singalila's Grasses and Sedges

Grasses and sedges form extensive landscape at the highest altitude below snowline. Principal features lie in the assemblage of many species together and in their intimate association. Principal grassland is of (in alpine and sub-alpine area) *Deyeuxia - Deschampsia* type.

Several species of grasses show strongly European character which are *Deschampsia caespitosa, Bromus mollis, Brachipodium sylvaticum, Poa annua, Festuca rubra, Holcus lanata, Dactylis glomerate* and few others.

Other notable grasses of the area are *Agrostis pilosula, A. nervosa, A. sikkimensis, Agropyron longearistatus, A. thoroldianum, Anthisteria hookeri, Anthoxanthon hookeri, Bromus asper, B. technorias, Danthonia cachmeryana, Deyeuxia scabrescens, D. pulchella, Poa annua, Elymus* sp. *Glyceria* sp., *Hierochloa* sp., *Kaeleria* sp., *Lolium* sp., *Oryzopsis* sp., *Phleum* sp., *Stipa* species and *Trisetum* sp.

All these facts presented in this work justify the region (Sikkim and Singalila mountains) as the richest floral and faunal zone and the nucleus of Natural Resources of the Great Himalayan Range.

BIBILIOGRAPHY

Abdulali H. 1965. The Birds of Andaman & Nicobar Islands. Jan. of Bom. Nat. His. Soc.

Alcock A.W. 1910. A catelogue of India Decapoda etc. Rec. Ind. Museum.

Annandale T.N. 1911. Fresh water sponges etc., Fauna of Bristish India.

Arunachal Forest News, 1980, 1987. Govt. of Arunachal

Biswas K. 1941. The flora of Akahills. In. For. Rec.

Blatter 1986. A note on the Organisation and activities of Andaman Forest Deptt.

Blandford W.T. 1970. Some notes on Reptila & Amphibia from Central India Jan. Asia. Soc. Bengal.

Bor N.L. 1938. A sketch of vegetation of Akahills. Ind. For. Rec.

Burkill H. 1924. The Botany of Abor hill expedition. Rec. Bot. Sur. India.

Champion H. & Sett S.K. 1968. Revised Classification of Forest types in India.

Chatterjee D. 1939. Studies on Endemic Flora of India & Burma. Journal Asia. Soc. Bengal.

C.M.F.R.I. Bull. 34 Sept. 1983. Mericulture potential of Andaman & Nicobar Islands.

Christopher S.R. 1933. Fauna of British India.

Day F. 1870. Of the Fishes of Andaman Island.

Elwes H.J. 1873. On the Orinithographical distribution of Asiatic Birds. Proc. Zoo. Soc. London.

Good R. 1947. Geography of Flowering Plants.

Jayakar K.C. 1974. Ecology & distribution of fresh water fishes, amphibia & reptila.

Kurup G.U. Mammals of India and Mammalian Geography of India.

Kurup G.U. 1966. Mammals of Assam and Adjoining Areas.

Kurz. S. 1875. Flora of Burma.

M&B Committee. Deptt. of Science and Tech. Preliminary inventory on potential area for Biosphe. Reserve.

M&B Committee. Deptt. of Environment. 1981. The Namdapha Biosphe. Reserve.

Mani M.S. 1968. Ecology and Biogeography of high altitude insects.

Mani M.S. 1974. Ecology and Biogeography of India.

Mc.Vean D.N. 1976. Reports on landuse in Andaman & Nicobar Islands.

Mukharjee A.K. 1982. Endangered animals of India.

Nair S.C. 1983. Natural Resource Conservation Dev. Report.

Prasad B. 1942. The Zoogeography of India.

Puri G.S. 1960. Indian Forest Ecology.

Rao G.C. et al 1985. The fauna of the Bay. Journal of And. Sanc. Society.

Rao V.M.K. 1983. Interim Report .of Biosphe. Reserve on North Andaman Is.

Sahni V.C. 1983. Bot. Expedition to Great Nicobar.

Sclater P.L. 1891. On the geographical distribution of birds - Ibis.

Smith M.H. 1934, 1938, 1943. The Geography of mammals.

Spate UHK 1951. India & Pakistan - A General & Regional Geography.

Tablbol G. 1939. Fauna of British India.

Wallace A.R. 1876. Geographical distribution of Animals.

Ward K. 1921. The distribution of Flora in S.E. Asia. Journal Ind. Bot. Soc.

Tikader B.K. 1984. Birds of Andaman & Nicobar Islands.

INDEX